Recent Advances in Biosensor Technology
(Volume 2)

Edited By

Vivek K. Chaturvedi

Department of Gastroenterology
Institute of Medical Sciences, Banaras
Hindu University, Varanasi, Uttar Pradesh
India

Dawesh P. Yadav

Department of Gastroenterology
Institute of Medical Sciences, Banaras Hindu University
Varanasi-221005
India

&

Mohan P. Singh

Centre of Biotechnology
Institute of Interdisciplinary Studies
University of Allahabad
Prayagraj-211002
India

Recent Advances in Biosensor Technology (Vol. 2)

Editors: Vivek K. Chaturvedi, Dawesh P. Yadav and Mohan P. Singh

ISSN (Online): 2972-4082

ISSN (Print): 2972-4074

ISBN (Online): 978-981-5136-41-8

ISBN (Print): 978-981-5136-42-5

ISBN (Paperback): 978-981-5136-43-2

Published by Bentham Science Publishers Pte. Ltd. Singapore. All Rights Reserved.

First published in 2023.

need for a court order if at any point you breach any terms of this License Agreement. In no event will any delay or failure by Bentham Science Publishers in enforcing your compliance with this License Agreement constitute a waiver of any of its rights.

3. You acknowledge that you have read this License Agreement, and agree to be bound by its terms and conditions. To the extent that any other terms and conditions presented on any website of Bentham Science Publishers conflict with, or are inconsistent with, the terms and conditions set out in this License Agreement, you acknowledge that the terms and conditions set out in this License Agreement shall prevail.

Bentham Science Publishers Pte. Ltd.
80 Robinson Road #02-00
Singapore 068898
Singapore
Email: subscriptions@benthamscience.net

BENTHAM
SCIENCE

CONTENTS

FOREWORD

In 1962, when the term "biosensor" was first introduced by Leland C. Clark and Champ Lyons by describing the use of an enzyme-based electrode for measuring glucose levels, people did not think that it would revolutionize the next-generation medicines. Biosensors are one of the groundbreaking and promising innovations. Therefore, detailed information on them is paramount for budding students and researchers. With great pleasure, I introduce the notable book "Recent Advances in Biosensor Technology." In this rapidly evolving era of science and technology, biosensors significantly transform the landscape of science and technology, healthcare and the environment to safety, toxicity and beyond. The area is witnessing an unprecedented expansion of biosensor applications, driven by breakthroughs in materials science, nanotechnology, and bioengineering. This book is a comprehensive compilation of modern research and cutting-edge developments in biosensor technology. It synthesizes the expertise and contributions of leading scientists, engineers, and researchers, offering a unique opportunity for readers to delve into the frontiers of this exciting domain. The chapters within this volume encompass a wide range of biosensor-related topics, providing an insightful exploration of various biosensing principles, design methodologies, and real-world applications. As we explore the contents of "Recent Advances in Biosensor Technology," we will have the opportunity to learn the most recent and significant advancements in nanomaterials, bio-molecular recognition elements, signal transduction mechanisms, and data analysis techniques. Whether a reader is a researcher, a student, or an industry professional, this book is a vital resource, bridging the gap between theory and practice. The authors' dedication has resulted in a compilation that will lead the future of biosensor technology and its societal impact, reflecting the limitless possibilities. I extend my heartfelt gratitude to the contributing authors for their dedication and commitment to sharing their knowledge with the scientific community. Their collective efforts have resulted in a remarkable compendium that will undoubtedly shape the future of biosensor technology and its impact on society. In conclusion, "Recent Advances in Biosensor Technology" proves the brilliance of human ingenuity and the boundless possibilities that lie ahead. I wish you to discover the same level of inspiration and enlightenment in this book I experienced.

Gyaneshwer Chaubey
Department of Zoology
Institute of Science
Banaras Hindu University
Varanasi-221005
India

PREFACE

Currently, there are urgent worldwide demands for biosensor development for human health. The development of biosensors by researchers is enhancing day by day by increasing their sensitivity and affinity towards the biomarkers. The role of nanotechnology is economical, reliable and a novel approach to diagnosis and therapeutics. Nanotechnology-based sensors are used for fast, ultra-sensitive and economical diagnostics. Nowadays, the coating of carbon-based and non-carbon-based nanomaterials such as graphene, nanoparticle or quantum dot is being used on the electrode surface to increase the signal amplification of biomarker load. Organic coating of Quantum dots changes nanoparticles into the hydrophilic compound to connect DNA, RNA, proteins, peptides and other molecules. In carbon-based nanostructures, graphene oxide is biocompatible and has the ability for immobilization of nanoparticles and single-strand DNA. CNTs play a key role in biosensors preparation that can detect target molecules in very little amounts. Silica-based electrochemical biosensors have a wide surface area, stability in critical thermal and chemical conditions and good compatibility with proteins. Nowadays, silver, gold and metallic nanoparticles have been extensively used in the field of virus detection owing to their unique optical, electrical properties, catalysis activity, and magnetic resonance imaging.

In this book, we covered several chapters regarding biosensors in the identification and characterization of natural bioactive compounds present in the food, pharmaceutical, agricultural, environmental, and industrial sectors. This book also demonstrates the types of biosensors (enzymatic, wearable and paper based) that integrate with the human body in the form of tattoo, gloves, and clothing that help in monitoring human health by calculating their daily routine. It is simple to do painless evaluations of bodily fluids using various biochemical markers such as spit, sweat, skin, and tears. This includes the 3d bioprinting of advanced bioinks for bone tissue regeneration; in addition, microfluidic technology, organ-on-a-chip, and electrospinning technology are used to produce biosensing products for the diagnosis and monitoring of living systems. This book comprises several new efficient biosensors that play a crucial role in the diagnosis revolution in neurodegenerative diseases, covid-19 pandemic, microorganisms, gastrointestinal diseases, diabetes management, as well as in agriculture. Thus, this book aims to broaden the readers' (academicians, students, and researchers) horizons and guide them in tailoring different biosensing techniques for specific diagnostic procedures. This book highlighted the latest development and the significant role of different new biosensors in future technology.

Vivek K. Chaturvedi
Department of Gastroenterology
Institute of Medical Sciences, Banaras
Hindu University, Varanasi, Uttar Pradesh
India

Dawesh P. Yadav
Department of Gastroenterology
Institute of Medical Sciences, Banaras Hindu University
Varanasi-221005, India

Mohan P. Singh
Centre of Biotechnology
Institute of Interdisciplinary Studies
University of Allahabad
Prayagraj-211002
India

List of Contributors

Abhay Dev Tripathi	School of Biochemical Engineering, Indian Institute of Technology (Banaras Hindu University), Varanasi-221005, India
Abha Mishra	School of Biochemical Engineering, Indian Institute of Technology (Banaras Hindu University), Varanasi-221005, India
Alma Tamunonengiofori Banigo	Department of Developmental Bio Engineering, Faculty of Science and Technology and TechMed Centre, University of Twente, Drienerlolaan 5, 7522 NB Enschede, The Netherlands
Anand Maurya	Institute of Medical Sciences, Faculty of Ayurveda, Department of Medicinal Chemistry, Banaras Hindu University, Varanasi, 221005, Uttar Pradesh, India
Anurag Kumar Singh	Centre of Experimental Medicine & Surgery, Institute of Medical Sciences, Banaras Hindu University, Varanasi 221005, Uttar Pradesh, India
Avanish Kumar Shrivastav	Department of Biotechnology, Delhi Technological University, Delhi, 110042, India
Bhaskar Sharma	Neurobiology Laboratory, Department of Anatomy, All India Institute of Medical Sciences, New Delhi, 110029, India
Chinedu Chamberlin Obasi	Department of Pharmaceutical Biotechnology, Institute of Pharmacy, Martin Luther University, Halle-Wittenberg, Germany
Divya Mishra	Centre of Bioinformatics, University of Allahabad, Prayagraj, Uttar Pradesh 211002, India
Dhitri Borah	Department of Zoology, Biswanath College, Biswanath Chariali,784176, Assam, India
Dawesh P. Yadav	Department of Gastroenterology , Institute of Medical Sciences, Banaras Hindu University , Varanasi-221005, India
Gaurav Mishra	Institute of Medical Sciences, Faculty of Ayurveda, Department of Medicinal Chemistry, Banaras Hindu University, Varanasi, 221005, Uttar Pradesh, India
Himani Yadav	Department of Botany, University of Delhi, Delhi-110007, India
Hemant Arya	Department of Biotechnology, Central University of Rajasthan, Ajmer, Rajasthan 305817, India
Marjan Talebi	Department of Pharmacognosy, School of Pharmacy, Shahid Beheshti University of Medical Sciences, Tehran, Iran
Manmath Kumar Nandi	Institute of Medical Sciences, Faculty of Ayurveda, Department of Medicinal Chemistry, Banaras Hindu University, Varanasi, 221005, Uttar Pradesh, India
Mohan P. Singh	Centre of Biotechnology, Institute of Interdisciplinary Studies, University of Allahabad, Prayagraj-211002, India
Mridula Chaturvedi	Amity Institute of Biotechnology, Amity University Noida, Uttar Pradesh, India

Neelesh Kumar Department of Biotechnology, Delhi Technological University, Delhi, 110042, India

Prem L. Uniyal Department of Botany, University of Delhi, Delhi-110007, India

Priti Giri Department of Botany, University of Delhi, Delhi-110007, India

Praveen Rai National Institute of Plant Genome Research, New Delhi, India

Ravi Kumar Goswami Department of Zoology, Hindu College, University of Delhi, New Delhi-110007, India

Rajendra Awasthi Department of Pharmaceutical Sciences, School of Health Sciences and Technology, University of Petroleum and Energy Studies (UPES), Energy Acres, Bidholi, Via-Prem Nagar, Dehradun – 248 007, Uttarakhand, India

Soumya Katiyar School of Biochemical Engineering, Indian Institute of Technology (Banaras Hindu University), Varanasi-221005, India

Sushil Kumar Dubey Centre of Biotechnology, University of Allahabad, Prayagraj, Uttar Pradesh 211002, India

Sujeet Singh Department of Biotechnology, Central University of Rajasthan, Ajmer, Rajasthan 305817, India

Tejveer Singh Translational Oncology Laboratory, Department of Zoology, Hansraj College, University of Delhi, New Delhi, Delhi-110067, India

Vivek Kumar Pandey Department of Pharmacology and Nutritional Sciences, University of Kentucky, Lexington, Kentucky, USA

Vivek K. Chaturvedi Department of Gastroenterology, Institute of Medical Sciences, Banaras Hindu University, Varanasi, Uttar Pradesh, India

LIST OF ABBREVIATIONS

ALG	Alginate
AM	Additive manufacturing
BBV	Bio-blood-vessel
BME	Biomedical engineering
BrCa	Breast cancer gene
CAD	Computer-aided design
CPF-127	Chitosan-g-pluronic F-127
CT	Computed tomography
3D	Three-dimension (al)
^{0}C	Degree Celsius
DCvC	Dynamic covalent chemistry
DN	Double network
DNA	Deoxyribonucleic acid
EB	Extrusion-based
EBB	Extrusion-based bioprinter
ECM	Extracellular matrix
GelMA	Gelatin methacrylamide
Gel-PEG-TA	Gelatin-Polyethylene glycol-tyramine
hMSCs	Human mesenchymal stem cells
HA	Hyaluronic acid
HAMA	Methacrylated hyaluronan
HA-TA	Hyaluronic acid-tyramine
HE	Heating element
HEPA	High efficiency particulate air
HMW	High molecular weight
HP	Hewlett-Packard
IR	Stereolithography Infrared Stereolithography
ICE	Ionic-covalent entanglement
IPNs	Interpenetrating networks
kDA	Kilo Daltons
kHz	One thousand hertz
LAB	Laser-assisted bioprinter

LED Light-emitting diode

LFS Low force stereolithography

MC Methylcellulose

MRI Magnetic resonance imaging

nHAp Hydroxyapatite nanoparticles

NM Nanometer

PEG Polyethylene glycol

PEGDA Polyethylene glycol diacrylate

PH Potential of hydrogen

RGD Arginyl-glycyl-aspartic acid

SLA Stereo lithograph apparatus

TE Tissue Engineering

UV Ultraviolet

UV-C Germicidal Ultraviolet-C

VdECM Vascular-tissue derived extracellular matrix

CHAPTER 1

The Emerging Role of Biosensors in the Identification, Characterization, and use of Natural Bioactive Compounds

Abhay Dev Tripathi[1,#], Vivek Kumar Pandey[2,#], Soumya Katiyar[1] and Abha Mishra[1,*]

[1] *School of Biochemical Engineering, Indian Institute of Technology (Banaras Hindu University), Varanasi-221005, India*

[2] *Department of Pharmacology and Nutritional Sciences, University of Kentucky, Lexington, Kentucky, USA*

Abstract: High specificity, less reagent requirement, swift response time, and high throughput screening make biosensors popular for detecting disease biomarkers, monitoring diseases, drug discovery, and other ecological applications such as the detection of environmental pollutants. Emerging health disorders demand targeted drugs with lower side effects in terms of advancing towards a low medicinal load. Active ingredients present in the different parts of plants and microorganisms have been used for several medical purposes for ages. Recent research has identified multiple promising natural bioactive compounds possessing the potential to mitigate several life-threatening diseases like Cancer, Diabetes, and Neurological disorders. Identifying such bioactive chemicals in the crude extract from various plant sources like leaves, roots, bark, fruits, and seeds, as well as in microbial extracts, is a tedious work that requires complex instrument setups, lengthy methods, and plenty of time, sometimes many years. The development and use of biosensors for natural bioactive moieties can overcome such problems and expedite the drug discovery process. This chapter provides a summary of the available biosensors for bioactive chemicals detection in extracts and fractions of organisms/plants, their types, design, and methods used for that purpose. Moreover, the chapter highlights the current use and the progress of the development of biosensors for identifying bioactive natural compounds.

Keywords: Biosensors, Cell-free biosensor, Cell-based biosensor, Nano-based biosensor, Phytochemicals.

* **Corresponding author Abha Mishra:** School of Biochemical Engineering, Indian Institute of Technology (Banaras Hindu University), Varanasi-221005, India; E-mail: abham.bce@itbhu.ac.in
# Shared co-first authorship contributed equally to this paper.

Vivek K. Chaturvedi, Dawesh P. Yadav and Mohan P. Singh (Eds.)

INTRODUCTION

Natural bioactive compounds are the chemicals produced by plants or the chemicals that are part of the plant's structure and physiology. These bioactive compounds are more commonly referred to as phytonutrients and can be part of the roots, stem, bark, leaves, flowers, and fruits of plants [1]. Bioactive compounds can be beneficial to health as they possess nutritional and medicinal values; on the other hand, they can be toxic and harmful to human health as well [2, 3] Major forms of bioactive compounds are classified as flavonoids, alkaloids, polyphenols, anthocyanins, pigments, vitamins, fatty acids, volatiles, and essential minerals [4]. In the past few decades, phytochemicals have been studied extensively for their medicinal efficacies and are well known for treating/lowering the risks of several diseases like malaria, measles, dysentery, constipation, stomachache, yellow fever, *etc.* Moreover, phytoconstituents are also well documented for their analgesic activities, antimicrobial activities, anti-inflammatory, anti-oxidative activities, anti-hyperglycemic activities, hepatoprotective and reno-protective activities [5– 8]. Recent research shows evidence of bioactive compounds regulating multiple signaling pathways during multiple metabolic disorders like diabetes and cancer [9– 11]. Flavonoids like Morin are reported to modulate insulin signaling, apoptotic pathways, ER stress, and the expression of glucose transporter proteins [7, 8, 12]. Moreover, several polyphenols like epigallocatechin gallate are reported to regulate MAPK kinases, p53, and other proteins involved in the progression of multiple cancers [13]. These examples demonstrate the potential of bioactive compounds to be exploited as therapeutic agents in several incurable diseases. So far, more than 10,000 phytochemicals have been identified, while very few (~150) have been evaluated in depth for their medicinal potential [14]. Moreover, a single plant and/or part of the plant may contain thousands of bioactive compounds, which could be beneficial for nutritional purposes as well as medicinal purposes. Identifying and characterizing these active compounds present in plants are not only complex but time taking too; developing specific biosensors to identify the specific type of the phytochemical may amplify the process. Biosensors can be very useful in the characterization of the bioactive compounds as well and can help in narrowing down the lead molecules out of numerous phytochemicals present in the crude extract.

Biosensors are devices that work based on biochemical reactions and measure the proportions of the analytes present or the end products of that biochemical reaction [15]. Biosensors require several components to work, including biocatalysts (enzymes), bioreceptor, transducers, signal amplifiers, display units, and the analyte that need to be detected [15, 16]. A transducer and a biosensing component are the conventional components of a biosensor. It is utilized for the detection of several components, including contaminants, a specific marker protein, microbial load, metabolites and numerous other compounds. It also has a wide range of uses in the medical field, food industries, and several other industries that demand accurate and reliable tests for screening samples [17]. A perfect example of a commonly used biosensor is Glucometer which is used to measure blood glucose levels.

This chapter explores the use of biosensors in identifying and characterizing phytochemicals which can be useful in drug discovery. The development of compatible and reliable biosensors for selecting bioactive compounds as lead molecules and identifying/isolating toxic plant chemicals can accelerate the speed of drug discovery and disease treatment.

CONSTRUCTING BIOSENSORS FOR DETECTING AND CHARACTERIZING PHYTOCHEMICALS

Development of the specific biosensors depends upon the information regarding the analyte to be detected, such as the chemical nature of the analyte, the concentration of the analyte, intermediated products, interfering species, and the type of matrix [18]. Identifying an end or intermediate product of the biochemical reaction that can be detected by the sensor of the biosensor is an important and tedious process in the development of biosensors for detecting bioactive plant chemicals [19]. Choosing a biosensor also depends upon the nature of the reaction and/or the type of sensors used to construct it. Below is the list of a few types of sensors or detectors used in developing biosensors:

Electrochemical sensors: These sensors are majorly based on the enzymatic catalysis that consumes or liberates electrons that can be captured by the sensors. Electrochemical sensors can be further classified into four types of sensors [20–22]:

a. *Amperometric Biosensors* (detect electrical current while keeping potential constant)
b. *Potentiometric Biosensors* (detect electrical potential)
c. *Impedimetric Biosensors* (detect electrical impedance)
d. *Voltammetric Biosensors* (detect current when applied to a specific voltage)

Immunosensors work based on the specificity of antibodies to their specific antigens.

Nucleic Acid biosensors work based on the interaction properties of DNA and RNAs to their complementary strands.

Optical biosensors work based on the optical activities of the analytes.

Calorimetric biosensors work based on thermal properties.

Magnetic biosensors utilize magnetic properties to detect analytes.

Fluorescent-based biosensors use synthetic chemistry and labeling of the analytes with fluorescent probes.

The selection of the sensors affects the reproducibility, accuracy, precision, and robustness of the results. Apart from that, the limit of detection and linearity are the other important factors to keep in mind while designing biosensors for identifying bioactive chemicals.

BIOSENSOR-BASED METHODS TO IDENTIFY AND CHARACTERIZE BIOACTIVE PLANT PRODUCTS

DNA probes, antibodies, enzymes, and cell receptors that interact with the analyte can all be used as biological sensing materials. A transducer, the above-mentioned biosensors (electrical, optical, *etc.*), converts the interaction signal into the detection patterns or signals that are measured. To brief the statement, molecules in the sensor interact with the reactants to produce an electrical signal that is directly proportional to the analyte's concentration. Based on this principle, biosensors use impedimetric, potentiometric, amperometric, *etc.*, to turn sensor data into a quantifiable signal (Fig. **1**) [23].

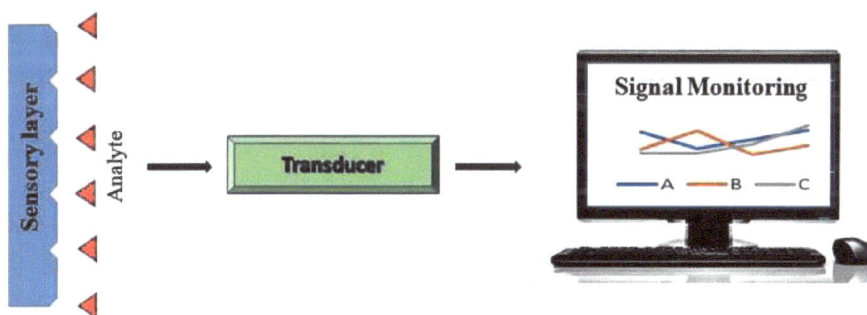

Fig. (1). Systematic components of a typical Biosensor.

At least one biological component serves as a detecting element in a biosensor. It might be a complete cell, an enzyme or protein, an aptamer, DNA or RNA, or an antibody. In comparison to other types of biosensors, enzyme-based biosensors have been used the most frequently as natural product detectors in industrial analysis. These biosensing interaction processes may be broken down into three categories for study: biosystems, bio affinity, and enzyme-based biocatalysts (Fig. 2) [24].

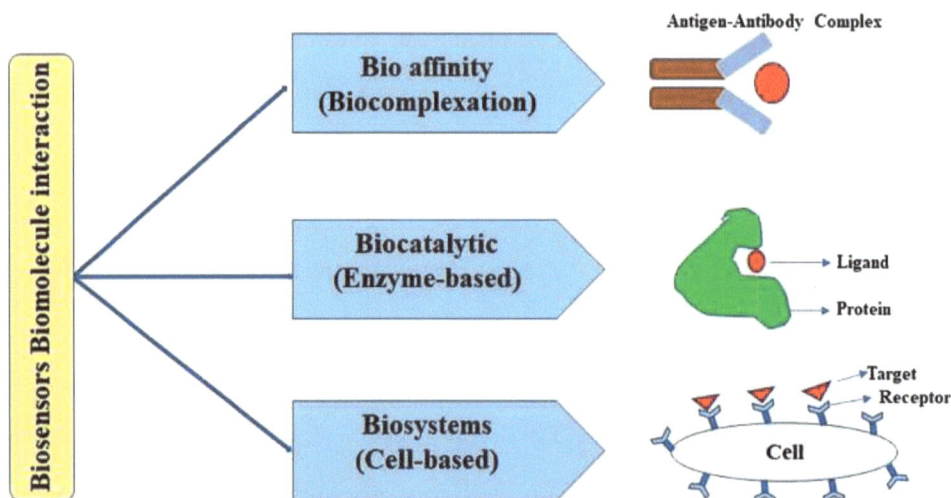

Fig. (2). Interactions types between Biomolecule (Analyte) and Biosensor.

Cell-free Biosensors

High-specificity and sensitivity biosensors have become important tools for prompt and accurate detection. Synthetic biology methods have been used for a wide range of substances. But unlike cell-free biosensors, many suggested cell-based biosensors have limitations in terms of shelf life, biosafety levels, and ethical regulatory concerns since they are based on live, genetically engineered organisms [25]. Additionally, a limiting step is the analyte's diffusion through the cell membrane [26]. Bio-affinity and Enzyme-based sensing can be categorized into Cell-free biosensor design, and various research studies developing biosensors for the detection of bioactive compounds are discussed. Schultz J *et al*. 1982 developed an affinity biosensor for the detection of glucose *via* immobilized concanavalin A protein [27]. A Quartz Crystal Microbalance affinity biosensor immobilized with desthiobiotin on a gold-layered surface for biotin detection was created by Masson *et al*. The basis of the biosensor working was the affinity between biotin and desthiobiotin [28]. For the aim of determining natural phytoestrogen polyphenols, Andreescu & Sadic, 2004 created an amperometric tyrosinase-based biosensor. This sensor can calculate the total phenolic concentration of phenolic estrogens found in nature, including quercetin, genistein and resveratrol [29]. Later in 2006, an optical tyrosinase-based biosensor for quantitative analysis of phenolic compounds was developed by Abdullah *et al*.

Chitosan film was used to immobilize the enzyme tyrosinase, and the enzymatic oxidation of phenol resulted in the production of o-quinone. This compound then combines with the reagent 3-methyl-benzothiazoline hydrazine to generate a maroon coloring detected using a spectrophotometer. For m-cresol, p-cresol, 4-chlorophenol, and phenol, this catalytic biosensor had LOD values of 1.0, 3.0, 0.9, and 1.0 M, respectively. Additionally, the sensor maintained its stability for almost two months [30].

Cell-based Biosensors

Compared to cell-free (molecule-based) receptors, using cell-based (entire cells) as sensing components has the advantages of being less expensive and more stable; that's because the requirement for purification is bypassed, cells can be grown in bulk quantities, and necessary co-factors are already present inside the cells. In a study conducted by Pfleger *et al.* for rapid screening and detection of mevalonate in an extracellular environment, a microbial-based biosensor has been employed utilizing *E. coli* mevalonate auxotroph [31]. For the screening of natural products with Quorum Sensing Inhibitors activity (QSIs), Ding *et al.* reported fluorescent-based whole-cell biosensors employing *C. violaceum* CV026 and *A. tumefaciens* A136. This biosensor was used to search a database of Traditional Chinese Medicine for QSIs operating on *P. fluorescens* as a spoiling microorganism agent that had minimal toxicity. Inhibition of the synthesis of long-chain and short-chain N-Acyl homoserine Lactones (AHLs) was shown by the loss of blue color and purple pigment, respectively [32]. To detect and monitor pamamycins, an antibiotic, Rebets *et al.* constructed a luminous whole-cell (*S. albus*) biosensor. The biosensor identified and regulated the antibiotic producers using a TetR transcriptional repressor [33]. Experimental research by Han *et al.* presented a FRET-based HeLa-C3 cells-based sensor for the detection of apoptosis-inducing ent-kaurene, a diterpenoid. To discover a prominent apoptotic agent in Chinese herbal extracts, researchers coupled a biosensor with a High-Speed Counter Current Chromatography method [34]. *S. mansoni*, a parasitic blood fluke, produces serotonin (5-HT), which interacts with cyclic AMP to control the motion of the parasite. Marchant *et al.* suggested using a luminous cAMP-biosensor in conjunction with co-expressed Sm.5HTRL receptor in mammalian cells to detect potential anti-schistosomal therapeutic compounds, such as protoberberines, aporphines, and tryptamines [35].

Nanotechnology-based Biosensors

Due to excellent conductivity, cytocompatibility and the high amount of molecule loading, nanomaterials provide a direct way for signal amplification in biosensor design [24]. Downsizing the biosensor to micro or nanoscale can improve the signal-to-noise ratio and allow for the use of limited sample amounts, which indirectly lowers the cost of the analysis. Additionally, once dimensions approach the nanometre scale, the surface-to-volume ratio of the sensor area rises, and because of this, the size of the detection electrodes and the targeted biomarkers becomes almost similar. Reduced non-specific binding and improved binding efficiency to the target molecule are the results of this [36]. Precise immobilization of the sensor components employing nanoparticles is achievable. The advantages of covalently coupling biomolecules to nanoparticles include stability, repeatability of surface modification, and a reduction in imprecise physical adsorption [37]. By employing immobilized laccase as the sensing component in the identification of phenolic natural products, Othman and Wollenberger were able to create an amperometric biosensor. To immobilize laccase, they utilized a coating of carboxyl carbon nanotubes (CNTs) on screen-printed carbon electrodes. The coating enhanced the amperometric responsiveness and made it possible to quickly identify phenolics [38]. A voltammetric tyrosinase-based sensor was constructed by Wee *et al.* to investigate phenolic metabolites in liquid setups. CNTs were used to create enzyme adsorption, participation, and crosslinking with tyrosinase molecules, *i.e.*, EAPC solution. The detection limit of the biosensor is significantly enhanced by using these nanostructures, which achieved 35 and 14nM LOD for phenol and catechol, respectively [39].

Magnetic nanoparticles improve the detection sensitivity by reducing the imprecise binding on the sensor region [40]. A nano-biosensor based upon mast cells was constructed by Xin *et al.* to identify (Botulinum Neurotoxin Type B) BoNT/B. For concentrated target absorption in test samples, researchers employed immobilized nano-magnetic beads layered with anti-BoNT/B polyclonal antibodies. The sensor's quick to monitor 100 pM BoNT/B [41]. A field-effect enzyme biosensor was constructed by Lin *et al.* to assess the quantities of proteins, H^+ ions, NH_2CONH_2 (urea), and $C_6H_{12}O_6$ in solutions. For this, the immobilization of enzymes was carried out in a magnetic powder using calcium-alginate microcubes. These magnetic beads are exposed to an external magnetic field thus, they serve as an enzyme carrier to fix the enzymes to the biosensor's surface. In such a setup, the assessed target contacts the biosensor's surface directly, enabling quantification of the target concentration [42].

Microfluid-based Biosensors

Microfluidic systems provide benefits such as quick functioning, requiring a minimal volume of samples and chemical compounds, increased accuracy, lesser power usage, and generating minimal waste. Additionally, the efficiency, controllability, and reproducibility of tests are also increased, and cross-contamination is decreased through biosensing assessment on microfluidic scales [24]. On a microfluidic system, Labroo and Cui developed an enzyme-graphen--based sensor for the consecutive monitoring of multiple metabolites. Every metabolite through this sensor is identified by a signal generated by the oxidase activity immobilized on the testing surface. In less than 2 minutes, the sensor identified xanthin, lactate, glucose, and cholesterol with a LOD of 0.3 M for each analyte [43]. A fluorescent-based microbial biosensor was created by Siedler *et al.* for the detection of secreted p-coumaric acid. Their method involved growing p-coumaric acid-producing cells in microfluidic droplets and introducing the monitoring bacteria into the droplet. Following injection, biosensor cell fluorescence signals were segregated and assessed [44]. A SAW-based (Surface Acoustic Wave) biosensor was utilized by Fournel *et al.* to detect the phycotoxin of okadaic acid in real-time. By incorporating a microfluidic module, the quality of the response pattern was enhanced, and the minimum okadaic acid detection concentration was reduced by elevating the flow volume and mass convection onto the bio-functional layer [45]. Table **1** summarizes the research progress made in the field of biosensor development for the detection of natural bioactive compounds.

Table 1. Studies for the development of the biosensor to detect natural bioactive compounds.

Natural Bioactive Detected	Bioreceptor	Detection Technique	Reference
Madindoline A	Human glycoprotein130 is covalently linked to the Fc region of the immunoglobulin (gp130-Fc-HA)	Surface Plasmon Resonance (SPR)	[46]
Phenolic compounds	Tyrosinase immobilized on a chitosan film	Colorimetry	[30]
Phytoestrogen polyphenols	Tyrosinase	Amperometry	[47]
Malbrancheamide and Tajixanthone	Fluorophore-labeled Calmodulin protein	Fluorescence	[48]
Ent-kaurene diterpenoids	HeLa–C3	FRET (Fluorescence Resonance Energy Transfer)	[49]
Okadaic acid	Inhibition of protein phosphatase 2A	Colorimetric	[50]
Domoic acid	AgNP (Silver Nanoparticles) as SERS substrate	Surface Enhanced Raman Scattering (SERS)	[51]
Okadaic acid	*In-situ* SERS analysis utilizing AgNP	SERS	[52]

(Table 1) cont.....

Natural Bioactive Detected	Bioreceptor	Detection Technique	Reference
Tetrodotoxin	Aptamer's structure conformational change	Fluorescent	[53]
Ciguatoxins	Sandwich ELISA	Fluorescent	[54]
Microcystins	Fluorescence quenching of graphene oxide (GO) by FRET between AuNPs and GO to detect microcystin-LR and microcystin-RR	FRET	[55]
Okadaic acid	SPR immunosensor conjugated with magnetic particles	SPR	[56]
p-coumaric acid	Bacterial cell	Fluorescence	[57]
Phenolic compounds	Tyrosinase is immobilized in Carbon nanotubes.	Voltammetry	[58]
Telomestatin	Biotin-labeled human telomeric oligonucleotide	SPR	[59]
Glutamate	Glutamate oxidase	Amperometry	[60]
Aflatoxin B1 (AFB1)	Anti-AFB1 antibodies conjugated to nanoparticles integrated into the gold chip	Amperometry	[61]
Benzoic acid derivatives	Multipeptide	Fluorescence	[62]
Phytoestrogen polyphenols	Tyrosinase	Amperometry	[63]
Volatile Organic Compounds	Odorant Binding Proteins	SPR	[64]

CONCLUSION

Natural bioactive compounds are mainly secondary metabolites that are synthesized by biological systems (plants, microorganisms, *etc.*) that have advanced uses, particularly in the food and medical sectors. Various studies have highlighted natural bioactive compounds as potential candidates for the treatment of various disease conditions like cancer, diabetes, and neurological disorders. Therefore, it becomes necessary to detect and quantify bioactive compounds. Meanwhile, the requirement of advanced techniques plays an important role in the easy and quick monitoring (detection) of natural bioactive compounds. The current scenario is focused on a novel technique, *i.e.*, in the development of biosensors. To increase the rate and limit of detection of bioactive, a promising screening strategy using biosensors as an analytical method can be useful. Biosensors can identify bioactive's inherent role in a biological system, measure their interaction with different molecules and quantify the concentration with ease. In a mixture of chemical products with identical structures (or *vice-versa*), biosensors can often precisely identify the targeted natural bioactive structures. Besides that, the substances employed in designing a biosensor's framework and selecting the suitable detecting method, affect the biosensor's effectiveness.

LIST OF ABBREVIATIONS

FRET	Fluorescence Resonance Energy Transfer
SAW-based	Surface Acoustic Wave based biosensor
BoNT/B	Botulinum Neurotoxin Type B
LOD	Limit of detection
MAPK	Mitogen-activated protein kinase
P53	Tumor protein P53
ER stress	Endoplasmic Reticulum Stress
QSI	Quorum Sensing Inhibitors activity
AHL	N-Acyl homoserine Lactones
Cyclic AMP	Cyclic adenosine monophosphate
DNA	Deoxyribonucleic acid
RNA	Ribonucleic acid
CNT	Carbon Nanotubes
EAPC	Enzyme Adsorption, Participation, and Crosslinking
SERS	Surface Enhanced Raman Scattering
SPR	Surface Plasmon Resonance

ACKNOWLEDGMENTS

The authors are thankful to the School of Biochemical Engineering, IIT (BHU) Varanasi. This work was financially supported by the Department of Biotechnology, India, under the DBT JRF Ph.D. program, for providing fellowship to author Abhay Dev Tripathi during the tenure of this study.

REFERENCES

[1]   J. Xiao, and W. Bai, "Bioactive phytochemicals", *Crit. Rev. Food Sci. Nutr.,* vol. 59, no. 6, pp. 827-829, 2019.
[http://dx.doi.org/10.1080/10408398.2019.1601848] [PMID: 31070480]

[2]   A.M. Bode, and Z. Dong, "Toxic phytochemicals and their potential risks for human cancer", *Cancer Prev. Res. (Phila.),* vol. 8, no. 1, pp. 1-8, 2015.
[http://dx.doi.org/10.1158/1940-6207.CAPR-14-0160] [PMID: 25348854]

[3]   Radha, M. Kumar, S. Puri, A. Pundir, S.P. Bangar, S. Changan, P. Choudhary, E. Parameswari, A. Alhariri, M.K. Samota, R.D. Damale, S. Singh, M.K. Berwal, S. Dhumal, A.G. Bhoite, M. Senapathy, A. Sharma, B. Bhushan, and M. Mekhemar, "Evaluation of nutritional, phytochemical, and mineral composition of selected medicinal plants for therapeutic uses from cold desert of Western Himalaya", *Plants,* vol. 10, no. 7, p. 1429, 2021.

[http://dx.doi.org/10.3390/plants10071429] [PMID: 34371632]

[4]   A.A. Vilas-Boas, M. Pintado, and A.L.S. Oliveira, "Natural bioactive compounds from food waste: Toxicity and safety concerns", *Foods,* vol. 10, no. 7, p. 1564, 2021.
[http://dx.doi.org/10.3390/foods10071564] [PMID: 34359434]

[5]   M. Mazani, Y. Mahmoodzadeh, M.M. Chinifroush Asl, S. Banaei, L. Rezagholizadeh, and A. Mohammadnia, "Renoprotective effects of the methanolic extract of Tanacetum parthenium against carbon tetrachloride-induced renal injury in rats", *Avicenna J. Phytomed.,* vol. 8, no. 4, pp. 370-379, 2018.
[PMID: 30377595]

[6]   V.K. Pandey, A. Mathur, M.F. Khan, and P. Kakkar, "Activation of PERK-eIF2α-ATF4 pathway contributes to diabetic hepatotoxicity: Attenuation of ER stress by Morin", *Cell. Signal.,* vol. 59, pp. 41-52, 2019.
[http://dx.doi.org/10.1016/j.cellsig.2019.03.008] [PMID: 30877037]

[7]   A. Mathur, V.K. Pandey, M.F. Khan, and P. Kakkar, "PHLPP1/Nrf2–Mdm2 axis induces renal apoptosis *via* influencing nucleo-cytoplasmic shuttling of FoxO1 during diabetic nephropathy", *Mol. Cell. Biochem.*, vol. 476, no. 10, pp. 3681-3699, 2021.
[http://dx.doi.org/10.1007/s11010-021-04177-3] [PMID: 34057658]

[8]   V. Kumar Pandey, A. Mathur, M. Fareed Khan, and P. Kakkar, "Endoplasmic reticulum stress induces degradation of glucose transporter proteins during hyperglycemic hepatotoxicity: Role of PERK-eIF2α-ATF4 axis", *Eur. J. Pharmacol.,* vol. 926, p. 175012, 2022.
[http://dx.doi.org/10.1016/j.ejphar.2022.175012]

[9]   S.L. Teoh, and S. Das, "Phytochemicals and their effective role in the treatment of diabetes mellitus: a short review", *Phytochem. Rev.,* vol. 17, no. 5, pp. 1111-1128, 2018.
[http://dx.doi.org/10.1007/s11101-018-9575-z]

[10]   V.K. Chaturvedi, N. Yadav, N.K. Rai, N.H.A. Ellah, R.A. Bohara, I.F. Rehan, G.E.S. Batiha, H.F. Hetta, and M.P. Singh, "Pleurotus sajor-caju-mediated synthesis of silver and gold nanoparticles active against colon cancer cell lines: a new era of herbonanoceutics", *Molecules,* vol. 25, no. 13, p. 3091, 2020.
[http://dx.doi.org/10.3390/molecules25133091] [PMID: 32645899]

[11]   V.K. Chaturvedi, S.N. Rai, N. Tabassum, N. Yadav, V. Singh, R.A. Bohara, and M.P. Singh, "Rapid eco-friendly synthesis, characterization, and cytotoxic study of trimetallic stable nanomedicine: A potential material for biomedical applications", *Biochem. Biophys. Rep.,* vol. 24, p. 100812, 2020.
[http://dx.doi.org/10.1016/j.bbrep.2020.100812] [PMID: 33083576]

[12]   P. Paoli, P. Cirri, A. Caselli, F. Ranaldi, G. Bruschi, A. Santi, and G. Camici, "The insulin-mimetic effect of Morin: A promising molecule in diabetes treatment", *Biochim. Biophys. Acta, Gen. Subj.,* vol. 1830, no. 4, pp. 3102-3111, 2013.
[http://dx.doi.org/10.1016/j.bbagen.2013.01.017] [PMID: 23352912]

[13]   R. Petric, C. Braicu, L. Raduly, N. Dragos, D. Dumitrascu, I. Berindan-Negoe, O. Zanoaga, and P. Monroig, "Phytochemicals modulate carcinogenic signaling pathways in breast and hormone-related cancers", *OncoTargets Ther.,* vol. 8, pp. 2053-2066, 2015.
[http://dx.doi.org/10.2147/OTT.S83597] [PMID: 26273208]

[14]   Y.J. Zhang, R.Y. Gan, S. Li, Y. Zhou, A.N. Li, D.P. Xu, and H.B. Li, "Antioxidant phytochemicals for the prevention and treatment of chronic diseases", *Molecules,* vol. 20, no. 12, pp. 21138-21156, 2015.
[http://dx.doi.org/10.3390/molecules201219753] [PMID: 26633317]

[15]   N. Bhalla, P. Jolly, N. Formisano, and P. Estrela, "Introduction to biosensors", *Essays Biochem.,* vol. 60, no. 1, pp. 1-8, 2016.
[http://dx.doi.org/10.1042/EBC20150001] [PMID: 27365030]

[16]   D. Garg, M. Singh, N. Verma, and Monika, "Review on recent advances in fabrication of enzymatic and chemical sensors for hypoxanthine", *Food Chem.,* vol. 375, p. 131839, 2022.

[http://dx.doi.org/10.1016/j.foodchem.2021.131839] [PMID: 34968951]

[17]   S. Singh, V. Kumar, D.S. Dhanjal, S. Datta, R. Prasad, and J. Singh, "Biological biosensors for monitoring and diagnosis". *Micro. biotech: basic research and applic.* Springerpp. 317-335, 2020.
[http://dx.doi.org/10.1007/978-981-15-2817-0_14]

[18]   F. Piroozmand, F. Mohammadipanah, and F. Faridbod, "Emerging biosensors in detection of natural products", *Synth. Syst. Biotechnol.,* vol. 5, no. 4, pp. 293-303, 2020.
[http://dx.doi.org/10.1016/j.synbio.2020.08.002] [PMID: 32954023]

[19]   C. Forzato, V. Vida, and F. Berti, "Biosensors and sensing systems for rapid analysis of phenolic compounds from plants: A comprehensive review", *Biosensors (Basel),* vol. 10, no. 9, p. 105, 2020.
[http://dx.doi.org/10.3390/bios10090105] [PMID: 32846992]

[20]   P. Mehrotra, "Biosensors and their applications – A review", *J. Oral Biol. Craniofac. Res.,* vol. 6, no. 2, pp. 153-159, 2016.
[http://dx.doi.org/10.1016/j.jobcr.2015.12.002] [PMID: 27195214]

[21]   B. Osa-Andrews, K. Tan, A. Sampson, and S. Iram, "Development of novel intramolecular FRET-based ABC transporter biosensors to identify new substrates and modulators", *Pharmaceutics,* vol. 10, no. 4, p. 186, 2018.
[http://dx.doi.org/10.3390/pharmaceutics10040186] [PMID: 30322148]

[22]   B. Fan, Q. Wang, W. Wu, Q. Zhou, D. Li, Z. Xu, L. Fu, J. Zhu, H. Karimi-Maleh, and C.T. Lin, "Electrochemical fingerprint biosensor for natural indigo dye yielding plants analysis", *Biosensors (Basel),* vol. 11, no. 5, p. 155, 2021.
[http://dx.doi.org/10.3390/bios11050155] [PMID: 34068869]

[23]   D. Grieshaber, R. MacKenzie, J. Vörös, and E. Reimhult, "Electrochemical Biosensors - Sensor Principles and Architectures", *Sensors (Basel),* vol. 8, no. 3, pp. 1400-1458, 2008.
[http://dx.doi.org/10.3390/s80314000] [PMID: 27879772]

[24]   F. Piroozmand, F. Mohammadipanah, and F. Faridbod, "Emerging biosensors in detection of natural products", *Synth. Syst. Biotechnol.,* vol. 5, no. 4, pp. 293-303, 2020.
[http://dx.doi.org/10.1016/j.synbio.2020.08.002] [PMID: 32954023]

[25]   A. Gräwe, A. Dreyer, T. Vornholt, U. Barteczko, L. Buchholz, G. Drews, U.L. Ho, M.E. Jackowski, M. Kracht, J. Lüders, T. Bleckwehl, L. Rositzka, M. Ruwe, M. Wittchen, P. Lutter, K. Müller, and J. Kalinowski, "A paper-based, cell-free biosensor system for the detection of heavy metals and date rape drugs", *PLoS One,* vol. 14, no. 3, p. e0210940, 2019.
[http://dx.doi.org/10.1371/journal.pone.0210940] [PMID: 30840628]

[26]   K. Yagi, "Applications of whole-cell bacterial sensors in biotechnology and environmental science", *Appl. Microbiol. Biotechnol.,* vol. 73, no. 6, pp. 1251-1258, 2007.
[http://dx.doi.org/10.1007/s00253-006-0718-6] [PMID: 17111136]

[27]   J.S. Schultz, S. Mansouri, and I.J. Goldstein, "Affinity sensor: a new technique for developing implantable sensors for glucose and other metabolites", *Diabetes Care,* vol. 5, no. 3, pp. 245-253, 1982.
[http://dx.doi.org/10.2337/diacare.5.3.245] [PMID: 6184210]

[28]   M. Masson, K. Yun, T. Haruyama, E. Kobatake, and M. Aizawa, "Quartz crystal microbalance bioaffinity sensor for biotin", *Anal. Chem.,* vol. 67, no. 13, pp. 2212-2215, 1995.
[http://dx.doi.org/10.1021/ac00109a047]

[29]   S. Andreescu, and O.A. Sadik, "Correlation of analyte structures with biosensor responses using the detection of phenolic estrogens as a model", *Anal. Chem.,* vol. 76, no. 3, pp. 552-560, 2004.
[http://dx.doi.org/10.1021/ac034480z] [PMID: 14750846]

[30]   J. Abdullah, M. Ahmad, N. Karuppiah, L.Y. Heng, and H. Sidek, "Immobilization of tyrosinase in chitosan film for an optical detection of phenol", *Sens. Actuators B Chem.,* vol. 114, no. 2, pp. 604-609, 2006.

[http://dx.doi.org/10.1016/j.snb.2005.06.019]

[31]   B.F. Pfleger, D.J. Pitera, J.D. Newman, V.J.J. Martin, and J.D. Keasling, "Microbial sensors for small molecules: Development of a mevalonate biosensor", *Metab. Eng.,* vol. 9, no. 1, pp. 30-38, 2007.
[http://dx.doi.org/10.1016/j.ymben.2006.08.002] [PMID: 17002894]

[32]   T. Ding, T. Li, and J. Li, "Identification of natural product compounds as quorum sensing inhibitors in *Pseudomonas fluorescens* P07 through virtual screening", *Bioorg. Med. Chem.,* vol. 26, no. 14, pp. 4088-4099, 2018.
[http://dx.doi.org/10.1016/j.bmc.2018.06.039] [PMID: 30100021]

[33]   Y. Rebets, S. Schmelz, O. Gromyko, S. Tistechok, L. Petzke, A. Scrima, and A. Luzhetskyy, "Design, development and application of whole-cell based antibiotic-specific biosensor", *Metab. Eng.,* vol. 47, pp. 263-270, 2018.
[http://dx.doi.org/10.1016/j.ymben.2018.03.019] [PMID: 29609044]

[34]   Q.B. Han, T. Yu, F. Lai, Y. Zhou, C. Feng, W.N. Wang, X.H. Fu, C.B.S. Lau, K.Q. Luo, H.X. Xu, H.D. Sun, K.P. Fung, and P.C. Leung, "Quick identification of apoptosis inducer from Isodon eriocalyx by a drug discovery platform composed of analytical high-speed counter-current chromatography and the fluorescence-based caspase-3 biosensor detection", *Talanta,* vol. 82, no. 4, pp. 1521-1527, 2010.
[http://dx.doi.org/10.1016/j.talanta.2010.07.036] [PMID: 20801367]

[35]   J.S. Marchant, W.W. Harding, and J.D. Chan, "Structure-activity profiling of alkaloid natural product pharmacophores against a Schistosoma serotonin receptor", *Int. J. Parasitol. Drugs Drug Resist.,* vol. 8, no. 3, pp. 550-558, 2018.
[http://dx.doi.org/10.1016/j.ijpddr.2018.09.001] [PMID: 30297303]

[36]   K.L. Adams, M. Puchades, and A.G. Ewing, "*In vitro* electrochemistry of biological systems", *Annu. Rev. Anal. Chem. (Palo Alto, Calif.),* vol. 1, no. 1, pp. 329-355, 2008.
[http://dx.doi.org/10.1146/annurev.anchem.1.031207.113038] [PMID: 20151038]

[37]   M. Holzinger, A. Le Goff, and S. Cosnier, "Nanomaterials for biosensing applications: a review", *Front Chem.,* vol. 2, p. 63, 2014.
[http://dx.doi.org/10.3389/fchem.2014.00063] [PMID: 25221775]

[38]   A.M. Othman, and U. Wollenberger, "Amperometric biosensor based on coupling aminated laccase to functionalized carbon nanotubes for phenolics detection", *Int. J. Biol. Macromol.,* vol. 153, pp. 855-864, 2020.
[http://dx.doi.org/10.1016/j.ijbiomac.2020.03.049] [PMID: 32165197]

[39]   Y. Wee, S. Park, Y.H. Kwon, Y. Ju, K.M. Yeon, and J. Kim, "Tyrosinase-immobilized CNT based biosensor for highly-sensitive detection of phenolic compounds", *Biosens. Bioelectron.,* vol. 132, pp. 279-285, 2019.
[http://dx.doi.org/10.1016/j.bios.2019.03.008] [PMID: 30884314]

[40]   M.J. Kwon, J. Lee, A.W. Wark, and H.J. Lee, "Nanoparticle-enhanced surface plasmon resonance detection of proteins at attomolar concentrations: comparing different nanoparticle shapes and sizes", *Anal. Chem.,* vol. 84, no. 3, pp. 1702-1707, 2012.
[http://dx.doi.org/10.1021/ac202957h] [PMID: 22224823]

[41]   W. Xin, W. Yao, X. Gao, Z. You, S. Gao, L. Kang, Q. Li, Y. Zhou, H. Yang, P. Jiang, and J. Wang, "Development of aequorin-based mast cell nanosensor for rapid identification of botulinum neurotoxin type B", *J. Biomed. Nanotechnol.,* vol. 10, no. 11, pp. 3318-3328, 2014.
[http://dx.doi.org/10.1166/jbn.2014.2026] [PMID: 26000390]

[42]   Y.H. Lin, C.P. Chu, C.F. Lin, H.H. Liao, H.H. Tsai, and Y.Z. Juang, "Extended-gate field-effect transistor packed in micro channel for glucose, urea and protein biomarker detection", *Biomed. Microdevices,* vol. 17, no. 6, p. 111, 2015.
[http://dx.doi.org/10.1007/s10544-015-0020-4] [PMID: 26553100]

[43]   P. Labroo, and Y. Cui, "Graphene nano-ink biosensor arrays on a microfluidic paper for multiplexed

detection of metabolites", *Anal. Chim. Acta,* vol. 813, pp. 90-96, 2014.
[http://dx.doi.org/10.1016/j.aca.2014.01.024] [PMID: 24528665]

[44] S. Siedler, N.K. Khatri, A. Zsohár, I. Kjærbølling, M. Vogt, P. Hammar, C.F. Nielsen, J. Marienhagen, M.O.A. Sommer, and H.N. Joensson, "Development of a bacterial biosensor for rapid screening of yeast p -coumaric acid production", *ACS Synth. Biol.,* vol. 6, no. 10, pp. 1860-1869, 2017.
[http://dx.doi.org/10.1021/acssynbio.7b00009] [PMID: 28532147]

[45] F. Fournel, E. Baco, M. Mamani-Matsuda, M. Degueil, B. Bennetau, D. Moynet, D. Mossalayi, L. Vellutini, J-P. Pillot, C. Dejous, and D. Rebière, "Love wave biosensor for real-time detection of okadaic acid as DSP phycotoxin", *Sens. Actuators B Chem.,* vol. 170, pp. 122-128, 2012.
[http://dx.doi.org/10.1016/j.snb.2011.02.056]

[46] E.M. Rezler, J. Seenisamy, S. Bashyam, M.Y. Kim, E. White, W.D. Wilson, and L.H. Hurley, "Telomestatin and diseleno sapphyrin bind selectively to two different forms of the human telomeric G-quadruplex structure", *J. Am. Chem. Soc.,* vol. 127, no. 26, pp. 9439-9447, 2005.
[http://dx.doi.org/10.1021/ja0505088] [PMID: 15984871]

[47] S. Andreescu, and O.A. Sadik, "Correlation of analyte structures with biosensor responses using the detection of phenolic estrogens as a model", *Anal. Chem.,* vol. 76, no. 3, pp. 552-560, 2004.
[http://dx.doi.org/10.1021/ac034480z] [PMID: 14750846]

[48] M. González-Andrade, M. Figueroa, R. Rodríguez-Sotres, R. Mata, and A. Sosa-Peinado, "An alternative assay to discover potential calmodulin inhibitors using a human fluorophore-labeled CaM protein", *Anal. Biochem.,* vol. 387, no. 1, pp. 64-70, 2009.
[http://dx.doi.org/10.1016/j.ab.2009.01.002] [PMID: 19185562]

[49] Q.B. Han, T. Yu, F. Lai, Y. Zhou, C. Feng, W.N. Wang, X.H. Fu, C.B.S. Lau, K.Q. Luo, H.X. Xu, H.D. Sun, K.P. Fung, and P.C. Leung, "Quick identification of apoptosis inducer from Isodon eriocalyx by a drug discovery platform composed of analytical high-speed counter-current chromatography and the fluorescence-based caspase-3 biosensor detection", *Talanta,* vol. 82, no. 4, pp. 1521-1527, 2010.
[http://dx.doi.org/10.1016/j.talanta.2010.07.036] [PMID: 20801367]

[50] A. Hayat, L. Barthelmebs, and J.L. Marty, "A simple colorimetric enzymatic-assay for okadaic acid detection based on the immobilization of protein phosphatase 2A in sol-gel", *Appl. Biochem. Biotechnol.,* vol. 166, no. 1, pp. 47-56, 2012.
[http://dx.doi.org/10.1007/s12010-011-9402-0] [PMID: 21984385]

[51] T.Y. Olson, A.M. Schwartzberg, J.L. Liu, and J.Z. Zhang, "Raman and surface-enhanced raman detection of domoic acid and saxitoxin", *Appl. Spectrosc.,* vol. 65, no. 2, pp. 159-164, 2011.
[http://dx.doi.org/10.1366/10-05910]

[52] S.C. Pinzaru, C. Müller, I. Ujević, M.M. Venter, V. Chis, and B. Glamuzina, "Lipophilic marine biotoxins SERS sensing in solutions and in mussel tissue", *Talanta,* vol. 187, pp. 47-58, 2018.
[http://dx.doi.org/10.1016/j.talanta.2018.05.006] [PMID: 29853065]

[53] Y. Lan, G. Qin, Y. Wei, C. Dong, and L. Wang, "Highly sensitive analysis of tetrodotoxin based on free-label fluorescence aptamer sensing system", *Spectrochim. Acta A Mol. Biomol. Spectrosc.,* vol. 219, pp. 411-418, 2019.
[http://dx.doi.org/10.1016/j.saa.2019.04.068] [PMID: 31059893]

[54] T. Tsumuraya, T. Sato, M. Hirama, and I. Fujii, "Highly sensitive and practical fluorescent sandwich ELISA for ciguatoxins", *Anal. Chem.,* vol. 90, no. 12, pp. 7318-7324, 2018.
[http://dx.doi.org/10.1021/acs.analchem.8b00519] [PMID: 29770692]

[55] Y. Shi, J. Wu, Y. Sun, Y. Zhang, Z. Wen, H. Dai, H. Wang, and Z. Li, "A graphene oxide based biosensor for microcystins detection by fluorescence resonance energy transfer", *Biosens. Bioelectron.,* vol. 38, no. 1, pp. 31-36, 2012.
[http://dx.doi.org/10.1016/j.bios.2012.04.053] [PMID: 22727517]

[56] D. Garibo, K. Campbell, A. Casanova, P. de la Iglesia, M. Fernández-Tejedor, J. Diogène, C.T. Elliott,

and M. Campàs, "SPR immunosensor for the detection of okadaic acid in mussels using magnetic particles as antibody carriers", *Sens. Actuators B Chem.,* vol. 190, pp. 822-828, 2014.
[http://dx.doi.org/10.1016/j.snb.2013.09.037]

[57]   S. Siedler, N.K. Khatri, A. Zsohár, I. Kjærbølling, M. Vogt, P. Hammar, C.F. Nielsen, J. Marienhagen, M.O.A. Sommer, and H.N. Joensson, "Development of a bacterial biosensor for rapid screening of yeast p -coumaric acid production", *ACS Synth. Biol.,* vol. 6, no. 10, pp. 1860-1869, 2017.
[http://dx.doi.org/10.1021/acssynbio.7b00009] [PMID: 28532147]

[58]   Y. Wee, S. Park, Y.H. Kwon, Y. Ju, K.M. Yeon, and J. Kim, "Tyrosinase-immobilized CNT based biosensor for highly-sensitive detection of phenolic compounds", *Biosens. Bioelectron.,* vol. 132, pp. 279-285, 2019.
[http://dx.doi.org/10.1016/j.bios.2019.03.008] [PMID: 30884314]

[59]   E.M. Rezler, J. Seenisamy, S. Bashyam, M.Y. Kim, E. White, W.D. Wilson, and L.H. Hurley, "Telomestatin and diseleno sapphyrin bind selectively to two different forms of the human telomeric G-quadruplex structure", *J. Am. Chem. Soc.,* vol. 127, no. 26, pp. 9439-9447, 2005.
[http://dx.doi.org/10.1021/ja0505088] [PMID: 15984871]

[60]   A.A. Karyakin, E.E. Karyakina, and L. Gorton, "Amperometric biosensor for glutamate using prussian blue-based "artificial peroxidase" as a transducer for hydrogen peroxide", *Anal. Chem.,* vol. 72, no. 7, pp. 1720-1723, 2000.
[http://dx.doi.org/10.1021/ac990801o] [PMID: 10763276]

[61]   H. Bhardwaj, G. Sumana, and C.A. Marquette, "A label-free ultrasensitive microfluidic surface Plasmon resonance biosensor for Aflatoxin B1 detection using nanoparticles integrated gold chip", *Food Chem.,* vol. 307, p. 125530, 2020.
[http://dx.doi.org/10.1016/j.foodchem.2019.125530] [PMID: 31639579]

[62]   S. Castaño-Cerezo, M. Fournié, P. Urban, J.L. Faulon, and G. Truan, "Development of a biosensor for detection of benzoic acid derivatives in saccharomyces cerevisiae", *Front. Bioeng. Biotechnol.,* vol. 7, p. 372, 2020.
[http://dx.doi.org/10.3389/fbioe.2019.00372] [PMID: 31970152]

[63]   S. Andreescu, and O.A. Sadik, "Correlation of analyte structures with biosensor responses using the detection of phenolic estrogens as a model", *Anal. Chem.,* vol. 76, no. 3, pp. 552-560, 2004.
[http://dx.doi.org/10.1021/ac034480z] [PMID: 14750846]

[64]   C. Hurot, S. Brenet, A. Buhot, E. Barou, C. Belloir, L. Briand, and Y. Hou, "Highly sensitive olfactory biosensors for the detection of volatile organic compounds by surface plasmon resonance imaging", *Biosens. Bioelectron.,* vol. 123, pp. 230-236, 2019.
[http://dx.doi.org/10.1016/j.bios.2018.08.072] [PMID: 30201334]

CHAPTER 2

Role of Wearable Biosensors in Healthcare

Himani Yadav[1], Bhaskar Sharma[2], Ravi Kumar Goswami[3], Avanish Kumar Shrivastav[4], Vivek K. Chaturvedi[5], Prem L. Uniyal[1] and Priti Giri[1,*]

[1] *Department of Botany, University of Delhi, Delhi-110007, India*

[2] *Neurobiology Laboratory, Department of Anatomy, All India Institute of Medical Sciences, New Delhi, 110029, India*

[3] *Department of Zoology, Hindu College, University of Delhi, New Delhi-110007, India*

[4] *Department of Biotechnology, Delhi Technological University, Delhi110042, India*

[5] *Department of Gastroenterology, Institute of Medical Sciences, Banaras Hindu University, Varanasi, Uttar Pradesh, India*

Abstract: These days, wearable biosensors are a very valuable tool for tracking the start of various acute and chronic diseases. Wearable biosensors (WBSs) are small, electrical devices that coordinate and collect sensations into the human body and can be present in the form of tattoos, gloves, clothing and inserts. WBSs are a flexible and practical tool for use in the healthcare industry, thanks to their ability to detect information, record it and estimate it accurately. WBSs help patients and doctors to communicate in both directions. It is simple to do painless evaluations of bodily fluids using various biochemical markers such as spit, sweat, skin, and tears. As the continuous state of capabilities of wearable and adaptable sensors continues to advance, the creation of new wearable gadgets that can fill the gap and handle the advantage of human well-being checking and clinical application is advancing. Blood is still the most crucial bio-liquid for assessing a person's health, even though more attention has been paid to other bodily fluids that are naturally secreted and severe functions that are similar to those of blood. There has been a lot of interest in the capacity of compact biosensing devices to identify the analyte in bio-liquids for the early detection of human well-being.

Keywords: Biomarkers, Gadgets, Healthcare, Wearable biosensors.

* **Corresponding author Priti Giri:** Department of Botany, University of Delhi, Delhi-110007, India; E-mail: giripriti06@gmail.com

Vivek K. Chaturvedi, Dawesh P. Yadav and Mohan P. Singh (Eds.)

INTRODUCTION

These days, the advancement and customization of wearable biosensors make them a very useful asset for monitoring the onset of various acute and chronic diseases. Wearable biosensors (WBSs) are compact electronic gadgets that coordinate and collect sensations into the human body and can be present in the form of tattoos, gloves, clothing and inserts. Detection of information, recording, and accurate estimation makes WBSs a versatile and convenient gadget for utilization in healthcare. Two-way communication is facilitated by WBSs among patients and doctors. Painless evaluation of human body liquids *via* different biochemical markers like spit, sweat, skin, and tears can achieve easily. With the development and coordinated application of material science, mechanical designing and advancements in remote correspondence, different wearable gadgets have been created and utilized for examining biomarkers to deliver appropriate medical care. With a maturing populace, the proof of sanitation and illness episodes has expanded. With the innovation and usability of WBSs, its market can ascend to USD 70 billion by 2025 [1].

A normal biosensor has two essential useful units, *i.e.*, a "biorecognition component or bioreceptor" (protein, immunizer, DNA, nucleic corrosive, peptide, and so on) and a physicochemical transducer, for example, piezoelectric, temperature, optical and electrochemical. The transducer is responsible for the change of a biorecognition occasion into quantifiable units, and the bioreceptor is liable for perceiving the objective analyte specifically. These gadgets were initially created and planned for single-use estimations, for instance, glucometer, glucose test strips, and glucowatch. Besides, the advancement in biosensors has prepared them for harmless checking in medical care and biomedical applications [2, 3].

WBSs can be ordered into biophysical sensors, movement state and biochemical sensors as per the various demands of the consumer. The movement sensors are utilized mainly to gauge actual human boundaries, for example, walking, rest, tremor and an assortment of long-haul data [4, 5]. With coordinated lab-on-chip innovation, researchers and lab labourers can exactly gauge biomarkers in organic liquids to screen for ailments and digestion [6 - 8]. Wearable biophysical sensors enchanted qualities to contact with skin to give an ongoing estimation of biophysical boundaries, for example, pulse and temperature, which have a huge breakthrough in medical care applications. The biophysical and movement state sensors are accessible and broadly utilized by buyers.

Among different types of biosensors, electrochemical-based sensors have shown maximum benefits because of their awareness, fast reaction, simplicity of development, and capacity to work with low utilization of force. Cathode detection plays an important part in the constitution of wearable sensors as detecting anodes also requires an innovative metal-based film cathode. Different headways have been accounted for by looking at new materials, such as the mixture of metallic nanoparticles, nanocomposite, carbon, and polymeric materials. Advancement in the miniature assembling of biosensors helps in improving the working boundaries of the detecting anodes [9]. Lately, huge endeavours have been made to achieve such sort of wearable sensors to perceive the different biomarkers that influence well-being. Finally, future viewpoints and difficulties are likewise examined for the enhancement WBSs in healthcare [10, 11].

WEARABLE DEVICES BASED ON DESIGN OR UTILITY

Customization of wearable biosensors in the administration of well-being has acquired huge consideration starting in the 21st century. Wearable gadgets can be named as per different wearable groups like wearable materials (socks, shirts and shoes), wearable gears (caps and glasses), and tactile gadgets (gloves and watches) for the assessment of well-being. With the coordinated scaled-down gadgets and headways in advances (microelectronics and remote correspondence), wearable biochemical sensors have profoundly implanted and turned into a vital piece of our lives [12 - 14].

Wrist-Mounted Wearable

Wearable Devices (WDs) are normally worn on the wrist. WDs are created with long battery life for checking the change in physiological boundaries to assess the crude signs into constant interpretable information. Earlier, wrist-mounted wearables were essentially accelerometer-based ones, for example, smartwatches, but nowadays, it has incorporated biometric detecting, *i.e.*, pedometer. Wristbands or smartwatches are utilized as harmless gadgets for monitoring human health [15].

Wristbands

Though watches and wristbands show resemblance, wristbands are famously sorted as wrist-worn wearable gadgets and are explicitly intended to follow human well-being. For instance, Jawbone made the UP4 band, which chips away at bio-impedance sensors to screen exercises, that is, strolling and following the capacity to record the dozing cycle. It similarly can catch signals, such as pulse, internal heat level, and galvanic skin reaction (GSR), utilizing different sensors (bio-impedance, tri-axis) situated on the inward side of the band. In any case, UP4 does not have a screen show, and the information can be perused through the cell phone application empowered in a cell phone. Other than UP4, there are different groups, for example, Huawei Talk band B3 and Fitbit. Based on the current market, medical care monitoring and checking equipment have a quickly growing revenue stream, especially the wristband market. Around 40 million gadget deals were anticipated in 2016, which were after the smart watches [14].

Wrist Watches

Today, one of the main wearable gadget types is smartwatches. In 2016, over 50 million units were sold, which is the second largest sale in the wearable tech-savvy gadget market. Normally, a smartwatch is a wellness GPS guide that assists clients with logging their everyday exercises, for example, naturally recording exercise times, following pulse, step counts, and calories consumed.

In the interim, checking arterial blood pressure (ABP) is promisingly an effective method for observing and controlling the common signs of cardiovascular disease (CVD) patients [16]. A smartwatch coordinated with a gyrator/accelerometer capability can be helpful in Parkinson's disease (PD) by recording tremors and balance in patients [17]. They surveyed the capacity of a smartwatch for the measurement of tremors in PD patients and assessment of clinical connection, its acknowledgement, and dependability as a checking device. Afterwards, it was viewed as promising. Moreover, planned savvy gadgets and fostered a calculation to identify atrial fibrillation (AF) from the information of the heart with the rate estimated using a Photoplethysmography (PPG) sensor and accelerometer [18]. The wrist-worn wearables are the primary supporters of making wearable items standard. The two primary sub-classes of wrist-worn (wearable gadgets worn on the wrist) smartwatches and wristbands right now address two unique client needs.

Wrist Patches

An adaptable and microfluidic-based fix framework for the ongoing examination of sweat tests was created [19]. This sensor is built on an adaptable plastic substrate coordinated with a unique winding channel microfluid inserted with particle-specific sensors. This framework interfaces the detecting part and is fit for examining sweat with a printed circuit board (PCB) innovation. The sensor can screen the grouping of particles (H^+, Na^+, K^+, Cl^-) and sweat rate, which further works with checking human physiological and clinical circumstances by sweat parameters. There is still a degree to work on the worldly goal of the sensors, which could empower the simplicity and high data in creation. To beat the shakiness concern, a fake atomically engraved polymer (MIP) blended from copolymerization response for cortisol screening was accounted [20]. As of now, different skin-connected wearable-fix or detecting stages are under the improvement stage, showing a moving concentration toward adaptable detecting.

Head-Mounted Devices

Visual tools with hands-free functionality and head-mounted tools are typically mounted on the user's head. Most research-related wearables fall into this category, which includes items like caps, spectacles, and helmets. Although these devices are currently used in modelling, imaging, and surgery, commercial head-mounted wearables appear to be less developed than wrist-worn ones [21]. The literature lists several head-mounted display systems that are currently used for navigation, education, simulation, surgery, and imaging [22]. These days, racers and participants in a variety of other sporting activities use helmets that have been specially fitted with temperature sensors.

Eyeglasses

Wearable systems (WSs) such as smart glasses are a type of head-mounted computer with a display capability. For instance, Nicholas created pulse-sensing smart glasses with a photoplethysmography (PPG) sensor on the nose pad that continually monitors heart rate [23]. The pulse-glass sensor was also compared with a lab electrocardiogram (ECG) system while a participant engaged in a variety of physical activities to cross-validate the heart rate data. The development of eyewear with a lactate biosensor incorporated into the nose pad allowed the simultaneous real-time monitoring of sweat lactate and potassium levels. An intrinsic benefit of these glasses is that they have switchable sensors. Numerous amperometric and potentiometric nose-bridge sensor stickers are available. One illustration is the interchangeability of the lactate bridge-pad sensor with the glucose bridge-pad sensor used for the monitoring of glucose in sweat. These wireless eyeglass sensor devices are completely integrated and may be expanded to simultaneously monitor electrolytes and metabolites in sweat fluid. Recon Jet, a sophisticated smart glass, for instance, uses the display to show information about the user's health status when they are cycling or jogging. Numerous smart eyeglasses have been developed, according to the literature, for a variety of uses, including tear biosensing to detect vitamins and minerals [24].

Cavitas

Mouthguards and contact lenses are only two examples of bodily cavities where wearable sensors can be placed. The word "cavity" has a Latin etymology called "Cavitas". These sensors extract data from biological fluid contained within a bodily cavity. There have been many cavitas sensors reported for detecting chemicals and biomolecules in tear fluid as well as transcutaneous gases in the mucosa of the eyelids [25].

With consistent and extended real-time monitoring of more than 5 hours, these sensors can detect glucose in artificial saliva. Similarly, to this, Kim *et al.*, 2014 showed that an integrated mouthguard with an enzyme-based biosensor for detecting salivary lactate and uric gave great selectivity and sensitivity. Since infants are unable to express their discomfort or health problems, the development of portable and non-invasive health monitoring systems for neonates is of tremendous interest [7].

Caps and Helmets

A helmet was created by a group of Danish scientists that is used to cure depression by reactivating body components associated with it and hastening patients' recovery through the delivery of mild electrical pulses to the brain. The FDA has cleared the helmet for use in treating depression by sending mild electrical impulses to a region of the brain associated with the condition. Two heads-up display-based systems that can reduce physiological disorders, including nausea, seizures, and body posture, have also been designed [7].

Smart Clothing/ E-Textiles

The basis for smart clothing (also known as "E-Textiles") is smart materials that can sense various environmental variables and react to stimuli like thermal, chemical, or mechanical changes. In Japan in 1989, the term "E-textiles" was first defined [26]. E-Textiles are a new, multidisciplinary field of wearable technology that have potential uses in safety, health, fitness, and medicine [27]. The development of conductive fabrics with embedded sensors on fabrics is a global endeavour involving several material sciences groups; nevertheless, this is outside the scope of the current review. These are fibres and filaments that can interact with the environment and the human body because they are made of conductive tools and clothing that is sewn or linked to them. E-textiles, which feature sensors like electrodes sewn into the fabric, serve as a nervous system to detect signals and are utilized to evaluate biofluids [28]. E-textiles are further categorized into three different forms [29].

Active E-textiles

Equipped with both an actuator function and a sensor device, they have a reactive character and can perceive external inputs from their environment.

Passive E-textiles

Material that can sense the environment or the user based on sensors incorporated;

Highly Intelligent E-textiles

Capable of detecting, responding to, and changing in specific situations.

E-textiles typically consist of three parts: a sensor, an actuator, and a controlling unit, and they are used to monitor physiological signals in humans as well as biomechanics and physical activity. The glucose and lactate-oxidase enzyme-coupled electrodes were integrated into the fabric to detect glucose and lactate [28]. A living material and a glove are also created by the same team and are integrated with hydrogel-elastomer hybrids into genetically modified bacteria that include genetic circuits to give the material a desired function. The different bacterial cell strains that were chemically generated in this experiment were contained in a hydrogel chamber. A diffusion process links the environment with the bacterial strain. When an inducer comes into contact with a bacterial sensor set to respond with fluorescence, it activates the response.

A stretchy printable protein-based cathode-based wearable electrochemical biosensor with the ability to recognize organophosphate (OP) nerve-specialist chemicals was planned [30]. The anode framework is printed using inks that withstand stress. The remote electronic surface was linked by a long serpentine relationship. A typical three-cathode framework, functional anodes, a reference Ag/AgCl-based anode, and a thumb-printed carbon cushion are included in the glove design. A hexoskin wearable vest that is suitable for monitoring heart rate(HR) and breath rate (BR) during routine activity was developed [31, 32]. A wearable biosensor based on conductive materials was also developed for BR detection using a capacitive detecting method. The typical position for a shirt to be worn is across the chest or middle, where the capacitance of two cathodes placed on the garments inside front and back can be used to estimate the breath cycle.

The development of wearable thermoelectric generators (TEG) for harvesting body heat typically, TEG was used to generate electrical energy from body heat, which in turn powered wearable technology [33].

For continuous and ongoing health monitoring, using an electrocardiogram (ECG), a smart shirt-based biosensor was developed to measure and speed increase heart signals. Wearable sensors are designed to be small enough to fit into shirts and use minimal power, which reduces the size of the battery. A flexible sifting approach in the intended shirt was created and tried to gain a reasonable ECG signal when running or engaging in an actual activity to drop the curio clamour from the cathodes made up of leading strands [34].

Chest-Mounted Devices

For parents or health professionals, assessing people's postural and falling impairments is crucial [35]. Two ready frameworks can be purchased for security checking: the Alter One clinical ready framework and the Life Alert Classic by Life Alert Emergency Response Inc. In these devices, a pendant and press button are synchronized, and pressing the button sends the message remotely to a different location. In addition, the Wellcore architecture makes use of advanced chip and accelerometer units to monitor postural growth. This device can distinguish between normal and abnormal body developments and discuss them further in a distant location. A chest-worn device called MyHaloTM by Halo Monitoring is also used to check the temperature, resting heart rate, and pulse. In conclusion, a device with a coordinated structure for a cell phone that is equipped with a balance sensor and activates pre-programmed dialling on crisis contact in the case of a fall will be extremely beneficial [35].

BIO-MULTIFUNCTIONAL SMART WEARABLE SENSORS (WSs)

One of the key components in the creation of WSs to mimic bio functions is the selection of nanomaterial with mechanical similarities. Examining several biological indicators, including physical, electrophysiological, and step capacities, as significant indicators of current deadly illnesses is one of the most important factors. The availability of wearable health monitoring devices in recent years has made it possible to identify these vital biological signals early. The wearable sensor's wear resistance and functionality can be improved by the availability of an optimal material. The types of bio-practical materials used in wearable sensors will be discussed in this section [36].

Self-Healing Flexible Wearable Sensors

Wearable medical devices are now constrained by their strength because of the ease with which biosensor components may be damaged, altering their capacity and further reducing their display, as well as their duration of usage and electrical characteristics [37]. The best bio-multifunctional wearable biosensors retain their electronic capabilities and have self-healing capacities to maintain their internal genuine characteristics in the event of slight micromechanical damage [38]. Wearable electronic devices placed on the skin should have the ability to repair themselves without external sensation (such as heat) to restore their mechanical and electrical connections. In light of guides and polymers, a few self-mending adaptable sensors have been investigated. A few of the self-mending polymeric materials, despite their rapid advancement, have been used in the creation of adaptable wearable technology. To achieve self-healing ability, a variety of composite materials that are packed with conductive particles or repair expert stacked casings are used.

Through the incorporation of ionic fluids into self-healing polymer channels, he and colleagues demonstrated the development of self-mending electronic sensor-based equipment. The tiny impact used in this design prevents the spilling of ionic fluids in a breaking condition. Self-healing electrochemical and wearable biosensors may be created by combining conductive ink that contains carbon and an acrylic stain cover [39, 40]. It showed a conductive composite that resembled elastic material consisting of inorganic micro nickel and a naturally occurring supramolecular polymeric molecule. This composite's electrical and mechanical self-mending mechanism is powered by hydrogen connections between cut surfaces. An adaptive sandwich main strain sensor was developed [41]. Most crucially, a few publications that may explain the advancements in the materials or composites used in wearable biosensors have now been widely disseminated [42].

Recently, hydrogels have distinguished themselves in state-of-the-art wearable sensors due to their mechanical characteristics [43]. The challenge, however, remains in creating a skin-like stretchy and conductive hydrogel with the desired synergistic features of stretchability, greater self-mending limit, and superior detecting execution. Through a two-step procedure, promoted the 3D organization of electro-conductive hydrogel and its use for tracking human movement. These hydrogels were used to create wearable sensors that tracked people's motions over time [44].

Despite hydrogel's tremendous advantages, its fragility and weakness provide two major obstacles to its continued use in wearable technology. Systems like double and interpenetrating organizations, such as double hydrogels, nanocomposite-based, and double cross-linked hydrogels, with strong mechanical characteristics and security in extreme conditions, can overcome these problems [45].

Recent years have seen a significant increase in interest in the presentation of dynamic polymer materials with a self-mending limit due to reversible bonds and dynamic connections. These polymeric materials are combined with synthetically viable particles to create an electronic skin that is thick, fluid, flexible, and capable of self-healing. Strangely, this substance can repair itself in various water environments (*i.e.*, deionized, seawater, very acidic, and basic arrangements). The reversible nature of dynamic materials has also been described in depth by several analysts, along with how they may be used in wearable electronic devices [46].

Biocompatible Wearable Sensor

Since wearable biosensors are directly connected to the human body, it is expected that they would not provide any additional health risks to life as we know it. Therefore, to prevent the occurrence of a safe reaction, the wearable biosensor needs to be biocompatible, making biocompatible materials the perfect choice for sophisticated wearable sensors.

Planning and organizing the use of nanoscale material in clinical and health-related concerns is now of utmost importance [47]. However, because of a series of obvious natural responses inside the human body, it is exceedingly difficult to conceive the relationship between a single chemical and the human biological system. To confirm and ensure the biocompatibility of material in a given application, a detailed plan and *in vivo* estimations are essential. It is important to use typical biocompatible polymers or materials for survival, such as cellulose, chitin, alginate, polydimethylsiloxane (PDMS), and polyurethane (PU), as they are benign. Recently, many biosensors that use these biomaterials have been extensively described. One of the brilliantly described materials with excellent elastic and conductive characteristics is chitosan [38].

Biodegradable Flexible Sensors

Recent publications claim that biodegradable technology has great promise for advancing high-level health monitoring and reducing the age of electronic trash. The negative effects of disease on human wellness have decreased as a result of these WB-based advancements. A sensitive and high-performance wearable strain device using biodegradable leading polymers was used for cardiovascular monitoring [48]. Greater responsiveness and faster reaction times enable the proposed sensor to be coordinated for continuous cardiovascular monitoring, such as the recording of blood-beat data. To avoid cautious mediation, it might also be applied in biological applications. A crucial method for handling tunes materials for biomedical applications is needed to deal with the design and construction of adaptable biosensors. The biosensor demonstrated excellent electrochemical action and security north of a couple of days in the location of dopamine and ascorbic corrosive with greater responsiveness due to the conductivity of micropatterns as the functioning cathode. In any event, there is a chance that an enzymatic reaction will attack these sensors [49].

Colourimetric-Based Wearable Sensors

In colourimetric-based detecting frameworks, the presence of development in a variety following compound responses among analyte and acknowledgement sites gives the quick location of the target atom. Colourimetric biosensors have been used to locate particles (H^+, Na^+, K^+, Ca^{2+}, Cl^-), single atoms, organisms, and proteins in the past. Lately, it is also possible to accomplish colourimetric localization utilizing synchronized high-quality mobile phone cameras and ordinary spectrophotometry equipment. The need for mechanical micropumps is also eliminated by combining this framework with materials to be used as material-based liquid handling stages gathering and examining. For example, a microfluidic-assembled colourimetric wearable biosensor coordinated concerning polyester/lycra material [50]. It possesses the characteristics of a liquid vehicle and is affected by the thickness and ratio of the two components. The pH sensor was created using a functionalized microfluidic texture combined with pH colour sensors and amazing liquid vehicle properties. An alternative technique also used is a microfluidic framework mounted on a cotton string and equipped with an LED-based locating system to work with sweat transport and measure the continual change in pH of sweat tests [51, 52].

It has been determined that the paper's presentation for the construction of wearable biosensors has the advantages of being lightweight, having a greater wicking rate, and being simple to functionalize. To detect sweat anions, including lactate, chloride, and bicarbonates, in cystic fibrosis patients presented a paper-skin fix-based screening technique [53]. To provide continuous skin contact, these sheets can be included in the adhesive skin fix. To measure and screen the water/drying out level and prevent decalcification of veneer independently, the paper-based devices also found interest in the location of pH levels in spit and sweat tests. Different colourimetric wearable devices constructed using plastic-based microfluidic technology have been taken into account when accounting for paper's capacity.

The development of a colourimetric localization technique to focus on the rate of sweat while exercising [54]. An evaluation of the sweat stream rate after a few cycles in video mode demonstrated the effectiveness of the tactic. PMMA-based microfluidic wearable biosensors integrated a variety of sweat to coordinate towards a working zone where pH may be calculated [51].

Electrochemical-Based Wearable Sensors

A greater understanding of the placement of electrochemical biosensors has also been made possible by advancements in nanotechnology, polymer science, and a combination of inorganic materials. Without microfluidic, a sweat analysis elective is made possible by combining an electrochemical biosensing stage with a pilocarpine iontophoresis system. However, due to the lack of sweat emission, these technologies are not appropriate for sweat investigation during resting time [54]. The sweat was directed towards a functional area by this sensor, which was coordinated with a particle-specific cathode for sodium-particle and a texture-based siphoning framework for assortment. Glass anode, the most practical material for wearable applications, was used in this device [55]. The cathode architecture was also included in a microfluidic read-out framework for continuous, uninterrupted monitoring. For testing ethanol, a T3 incorporating cryogels and screen-printed cathodes were used. The location system employed an amperometric discovery for liquor assurance that used the enzymatic oxidation of substrate by liquor oxidase. To identify glucose and lactate in sweat testing, the development of a skin-mounted wearable device was described [56].

Biomarkers in Biofluids

Biomedical devices and wearable biosensors have shown their importance in drug metabolites, the location of biomarkers and chemicals in various organic liquids and grids for years and years earlier due to their inherent attributes and potential application [57].

Spit-Based Wearable Biosensors

Because spit contains a variety of disease-flagging biomarkers that reflect a person's overall health, interest in spit as a liquid indicator of symptoms has increased significantly over the past few years. Some of the justifications for this acceptance include the existence of many disease-flagging salivary biomarkers that accurately depict well and ill states in people, as well as the testing advantages over blood examination. Spit is a reflection of human health because several natural indicators diffuse from the circulatory system *via* transcellular/paracellular mechanisms. These biosensors provide an alternative route to a blood test for examining human metabolites, such as chemicals and proteins. The parotid organ produces spit, a highly complex biofluid with a high protein content that contains a few key components such as drug metabolites, substances, microbial verdure, and chemicals. There are not many publications on wearable spit biosensors, perhaps because of the biofouling caused by the high protein content of saliva and the poor grouping of the analyte to be identified, even though these biomarkers have already been used in diagnosis. Nevertheless, wearable oral biosensing devices may provide a seductive and simple method for obtaining dynamic chemical data from the spit. Wearable salivary biosensing has emerged as a viable technology with the recent improvement in salivary diagnostics [58].

Tear-Based Wearable Biosensors

Human tears are a major and complex natural liquid, similar to spit and perspiration, made up of various proteins, electrolytes, metabolites, and more than

98% water. Tears in various forms may be useful for analysing human metabolites. The development of tear-based technology has received noteworthy attention over the past century, yet this sector is capable of looking into wearable technology for tear inspection. Contact focal points, which are in direct contact with the basal tears, are an appropriate framework for gathering tears without harming the eye. With the essential biosensing frameworks, they can be easily coordinated. As a result of the connection between glucose and concanavalin. A (or phenylboronic corrosive) subsidiaries by optical estimations, contact focal point-based detecting stages were initially developed for glucose testing in a tear. A mobile phone is used to record the intelligent force of the focal point concerning the adjustment of tear glucose level. This strategy might provide a quick and sensitive location for glucose production. The tear-assembled biosensor is typically used to assess glucose levels, but there is a lot of room for the safe detection of other physiologically important indicators. The analyte whose concentration in tear liquid is closely similar to that in the blood can be consolidated, broadening the application for additional analytes. On the other hand, a specific test for tear-based wearable biosensors is the identification of an acceptable power source and its appropriate size. Biofuel cells (BFCs) are thought to be a workable solution for locally generating power to overcome this. The finest examples of a biofuel or analyte to recognize include ascorbate, lysozymes, and pyruvate, aside from glucose. As a result, the researchers developed a successful BFC design that was tested on human tears and was based on nanostructured microelectrodes wrapped in gold nanowires. However, this approach was unsuccessful in establishing the relationship with blood concentrations, evaluating the size, and determining the opposing effects [59].

Sweat-Based Wearable Biosensors

Sweat organs, which are carried throughout the body by sweat, are a significant biofluid that can provide wearable and painless detection of a patient's health status [60]. Additionally, the skin's physiology and the availability of shifting metabolites, electrolytes (Na^+, K^+, $NH4^+$, $Ca2^+$), chemicals, and natural poisons make them the best places to test for and diagnose metabolic [61]. These wearable sensors are incredibly practical and can continuously monitor well-being. The presence of biomarkers in sweat may be of great relevance for non-invasively assessing an individual's health, such as hydration level and disease state (like diabetes, or cystic fibrosis). The necessity for blood testing can be eliminated further with *in-situ*, non-harmful sweat analysis at the epidermal layer. However, additional permission is also anticipated to evaluate the clinical use of sweat as a biofluid indicator [60].

A thorough understanding of the science of sweat and the metabolite transport system, along with advancements in sweat testing and localization techniques, could hasten the practical application of sweat-based symptomatic open doors. In addition to providing superior skin contact to material-based biosensors, epidermal wearable technology has a shorter lifespan [62].

A remotely powered fix-type wearable sensor has improved, in light of glue radio-recurrence identification (RFID) [63]. This sensor can mimic human skin and display the biomarkers that are measured during a perspiration test. According to this design, an electrical circuit with an RFID was built on a glue fix and used for potentiometric measurements of analytes in sweat testing. The findings were remotely viewable on an Android mobile application. This epidermal patch provides simultaneous health monitoring and L-lactate detection in a sweat test using Prussian blue (PB). The combination of all components (tactile component, microcontroller and remote communication module on a single chip was a very well-organized design constraint.

Implantable and Subcutaneous Wearable Biosensors

The development of devices for subcutaneous-wearing intercellular fluid (ICF) testing has advanced recently. The best target for the development of compact wearable devices for continual monitoring of interstitial fluid (ISF), particularly the components in it, are different significant ISF components such as particles (Na^+, K^+, Cl^-), metabolites (lactate, glucose, chemicals), and so on [64]. The combination of developments in small dialysis tests and microneedles attracted a lot of attention in the field. A coaxial microfluidic test made of polymeric material with an optimal eliminated pore size and placed out in close contact with the tissue is a common component of small-scale dialysis procedures. An additional risk factor for analyte dispersion through the layer inside the test is the analyte focus angles across the film and tissue. A predetermined volume of dialysate is either collected for downstream analysis to determine the analyte fixation or ongoing web-based techniques, such as HPLC and LC-MS, are used to determine the analyte focus from the dialysate [65]. The current limitations of small dialysis procedures are due to ongoing data points or the demand for large logical devices for assessment. To measure lactate and glucose by an embedded test with an intended time objective of the 30s continuously, it was straightforward to do delayed high worldly examination using tubing rather than the bottles used to store dialysate.

Additionally, this method addresses the issue of examination variety deferral. On the other hand, the development of microneedles relies on the use of a variety of tiny, precisely measured needles made of rigid materials like silicon, biocompatible polymers, or hard plastics. Combining microneedle technology allowed for the simultaneous use of *in-situ* medication delivery and analyte placement in interstitial fluids (ISFs). Wearable examples have made use of microneedles-based techniques for biomarker checking and performing the function of authorized medication conveyance.

There are still certain gaps in these devices, despite their commercial and clinical successes, since a finger stick blood test must still be done when using self-implantable biosensor devices. Currently, a lot of effort needs to be made to establish the reliability, strength, and biocompatibility of the material selected for wearable, implantable biosensors with a longer usable lifespan (up to a few months or years) and more varied applications [66].

CONCLUSION AND FUTURE PERSPECTIVES

It is crucial to accurately analyse the factors that influence human well-being and the organic signs that show a person's illness. Since the emergence of lab-on-chip computing devices, wearable bioelectronics with human skin and tissue interfaces have been developed. Working on the security, durability, and dependability of the test equipment is encouragingly crucial while designing a flexible wearable device. In the event of persistent illness and its uninterrupted constant well-being checking, there is a chance to obtain superior clinical perspectives and logical location. It is also stated that there is a need for wearable technology that is swift, sturdy, and simple to wear. In the current audit, we thoroughly looked at wearable sensors and their different types with their inherent characteristics, focusing particularly on the wearable-based design and utility of sensors. Different applications have been looked at, but high accuracy, biocompatibility, higher responsiveness, exactness, power, and security of wearable technology are still needed.

The development of new wearable devices that can cover the gap and handle the advantage of human well-being checking and clinical application is advancing as the ongoing status of the capabilities of wearable and adaptable sensors is increasing. Compact biosensing devices' ability to determine the analyte in bio-liquids for the early detection of human well-being has drawn a lot of attention. The most important bio-liquid for determining a person's health is still their blood, although the greater focus has been placed on other bodily fluids that are naturally released and have a similar function to blood. It is possible to establish a direct link between the target analyte and organic liquids. As there are painless ways to collect and dissect samples, body fluids like perspiration, tears, and spittle are studied in this way.

The current condition of the craft of wearable biosensing swims around, displaying the evidence of the notion of wearable gadgets for the assurance of various biomarkers, notwithstanding significant advancement and improvement in wearable biosensors/gadgets. In any event, not many steps have been taken in this sector toward practical applications and commercialization. Strongness, accuracy, exactness, security, consistency, and other parameters are taken into consideration. Wearable biosensors encountered various significant problems and innovation gaps related to the degree. Overcoming these obstacles is essential for successful development and widespread commercialization. Biosensors have undergone a significant transformation concerning awareness, vigour, security, and portability, thanks to advancements in microfluidic mixing and biosensor scaling down. Future planning and development of unique wearable technology will be influenced by a variety of the aforementioned boundaries rather than just one. The benefits of visual location and coordinated portable-based readout architecture are provided by the colourimetric biosensor. Additionally, electrochemical approaches are associated with increased awareness. In the current audit, we looked at the importance of wearable sensors that are microfluidic-based and have more advanced implementation. The variety of tests, minimization of the test volume, and specificity of the delivery of collected instances like sweat, spit, ISF, and tears are important for an effective approach concerning the sensor's dynamic detecting region. These will contribute to the development of wearable biosensor technology by providing characteristics of microfluidic stages.

The position of a wide range of analytes, such as particles (Na^+, K^+, NH_4^+, Ca^{2+}), and human metabolites (lactate, glucose), in various organic liquids has been accounted for by several high-level and innovative wearable biosensing systems/gadgets. Using both enzymatic and non-enzymatic catalysis, these resemble blood. Wearable technology has benefited from advancements in testing methods, accessibility of adaptable and biodegradable materials, and remote communications. This innovation indirectly focuses on the dependability of these sensors, the capability of analyte checking, and wearability.

By the way, we have faith in the power of invention and are waiting for imaginative trial and error and methods with increased sufficiency. Worldview-changing developments typically arrive quickly and are given attention with the upgrades and globalization trends in the biosensor market. It would be a significant triumph for the medical care industry and patient life shortly, should these new patterns gain commercial success.

LIST OF ABBREVIATIONS

ABP	Arterial Blood Pressure
AF	Atrial Fibrillation
BFCs	Biofuel Cells
BR	Breath Rate
CVD	Cardiovascular Disease
ECG	Electrocardiogram
FDA	Food and Drug Administration
GSR	Galvanic Skin Reaction
HR	Heart Rate
ICF	Intercellular Fluid
ISF	Interstitial Fluid
MIP	Molecularly Imprinted Polymer
OP	Organophosphate
PB	Prussian Blue
PCB	Printed Circuit Board
PD	Parkinson's Disease
PDMS	Polydimethylsiloxane
PMMA	Polymethylmethacrylate
PPG	Photoplethysmography
PU	Polyurethane
RFID	Radio-recurrence identification
TEG	Thermoelectric Generators
WBSs	Wearable Biosensors
WD	Wearable Devices
WSs	Wearable Systems

ACKNOWLEDGEMENTS

V.K.C. gratefully acknowledges the Department of Health Research (DHR), Govt. of India, for support through the Young Scientist Fellowship Grant R.12014/56/2022-HR.

REFERENCES

[1]   S. Ajami, and F. Teimouri, "Features and application of wearable biosensors in medical care", *J. Res. Med. Sci.,* vol. 20, no. 12, pp. 1208-1215, 2015.
       [http://dx.doi.org/10.4103/1735-1995.172991] [PMID: 26958058]

[2]   A. Sharma, A. Chandra Singh, G. Bacher, and S. Bhand, "Recent advances in aptamer-based biosensors for detection of antibiotic residues", *Aptamers Synth. Antibodies,* vol. 2, pp. 43-54, 2016.

[3]   A. Sharma, K. Goud, A. Hayat, S. Bhand, and J. Marty, "Recent advances in electrochemical-based sensing platforms for aflatoxins detection", *Chemosensors (Basel),* vol. 5, no. 1, p. 1, 2016.
        [http://dx.doi.org/10.3390/chemosensors5010001]

[4]   W. Song, B. Gan, T. Jiang, Y. Zhang, A. Yu, H. Yuan, N. Chen, C. Sun, and Z.L. Wang, "Nanopillar arrayed triboelectric nanogenerator as a self-powered sensitive sensor for a sleep monitoring system", *ACS Nano,* vol. 10, no. 8, pp. 8097-8103, 2016.
        [http://dx.doi.org/10.1021/acsnano.6b04344] [PMID: 27494273]

[5]   S. Xia, S. Song, F. Jia, and G. Gao, "A flexible, adhesive and self-healable hydrogel-based wearable strain sensor for human motion and physiological signal monitoring", *J. Mater. Chem. B Mater. Biol. Med.,* vol. 7, no. 30, pp. 4638-4648, 2019.
        [http://dx.doi.org/10.1039/C9TB01039D] [PMID: 31364689]

[6]   J. Choi, A.J. Bandodkar, J.T. Reeder, T.R. Ray, A. Turnquist, S.B. Kim, N. Nyberg, A. Hourlier-Fargette, J.B. Model, A.J. Aranyosi, S. Xu, R. Ghaffari, and J.A. Rogers, "Soft, skin-integrated multifunctional microfluidic systems for accurate colorimetric analysis of sweat biomarkers and temperature", *ACS Sens.,* vol. 4, no. 2, pp. 379-388, 2019.
        [http://dx.doi.org/10.1021/acssensors.8b01218] [PMID: 30707572]

[7]   J. Kim, G. Valdés-Ramírez, A.J. Bandodkar, W. Jia, A.G. Martinez, J. Ramírez, P. Mercier, and J. Wang, "Non-invasive mouthguard biosensor for continuous salivary monitoring of metabolites", *Analyst (Lond.),* vol. 139, no. 7, pp. 1632-1636, 2014.
        [http://dx.doi.org/10.1039/C3AN02359A] [PMID: 24496180]

[8]   X. Li, C. Zhao, and X. Liu, "A paper-based microfluidic biosensor integrating zinc oxide nanowires for electrochemical glucose detection", *Microsyst. Nanoeng.,* vol. 1, no. 1, p. 15014, 2015.
        [http://dx.doi.org/10.1038/micronano.2015.14]

[9]   S. Kabiri Ameri, R. Ho, H. Jang, L. Tao, Y. Wang, L. Wang, D.M. Schnyer, D. Akinwande, and N. Lu, "Graphene electronic tattoo sensors", *ACS Nano,* vol. 11, no. 8, pp. 7634-7641, 2017.
        [http://dx.doi.org/10.1021/acsnano.7b02182] [PMID: 28719739]

[10]   Y.S. Rim, S.H. Bae, H. Chen, J.L. Yang, J. Kim, A.M. Andrews, P.S. Weiss, Y. Yang, and H.R. Tseng, "Printable ultrathin metal oxide semiconductor-based conformal biosensors", *ACS Nano,* vol. 9, no. 12, pp. 12174-12181, 2015.
        [http://dx.doi.org/10.1021/acsnano.5b05325] [PMID: 26498319]

[11]   J. Xiong, P. Cui, X. Chen, J. Wang, K. Parida, M.F. Lin, and P.S. Lee, "Skin-touch-actuated textile-based triboelectric nanogenerator with black phosphorus for durable biomechanical energy harvesting", *Nat. Commun.,* vol. 9, no. 1, p. 4280, 2018.
        [http://dx.doi.org/10.1038/s41467-018-06759-0] [PMID: 30323200]

[12]   K. Guk, G. Han, J. Lim, K. Jeong, T. Kang, E.K. Lim, and J. Jung, "Evolution of wearable devices with real-time disease monitoring for personalized healthcare", *Nanomaterials (Basel),* vol. 9, no. 6, p. 813, 2019.
        [http://dx.doi.org/10.3390/nano9060813] [PMID: 31146479]

[13]   K. Kaewkannate K., and S. Kim, "A comparison of wearable fitness devices", *BMC Public Health,* vol. 16, no. 433, 2016.

[14]   S. Seneviratne, Y. Hu, T. Nguyen, G. Lan, S. Khalifa, K. Thilakarathna, M. Hassan, and A. Seneviratne, "A survey of wearable devices and challenges IEEE", *IEEE Commun. Surv. Tutor.,* vol. 19, no. 4, pp. 2573-2620, 2017.
        [http://dx.doi.org/10.1109/COMST.2017.2731979]

[15]   A. Kamišalić, I. Fister Jr, M. Turkanović, and S. Karakatič, "Sensors and functionalities of non-invasive wrist-wearable devices: A review", *Sensors (Basel),* vol. 18, no. 6, p. 1714, 2018.

[http://dx.doi.org/10.3390/s18061714] [PMID: 29799504]

[16]   S. Rastegar, H. GholamHosseini, and A. Lowe, "Non-invasive continuous blood pressure monitoring systems: current and proposed technology issues and challenges", *Physical and Engineering Sciences in Medicine,* vol. 43, no. 1, pp. 11-28, 2020.
[http://dx.doi.org/10.1007/s13246-019-00813-x]

[17]   R. López-Blanco, M.A. Velasco, A. Méndez-Guerrero, J.P. Romero, M.D. del Castillo, J.I. Serrano, E. Rocon, and J. Benito-León, "Smartwatch for the analysis of rest tremor in patients with Parkinson's disease", *J. Neurol. Sci.,* vol. 401, pp. 37-42, 2019.
[http://dx.doi.org/10.1016/j.jns.2019.04.011] [PMID: 31005763]

[18]   G.H. Tison, J.M. Sanchez, B. Ballinger, A. Singh, J.E. Olgin, M.J. Pletcher, E. Vittinghoff, E.S. Lee, S.M. Fan, R.A. Gladstone, C. Mikell, N. Sohoni, J. Hsieh, and G.M. Marcus, "Passive detection of atrial fibrillation using a commercially available smartwatch", *JAMA Cardiol.,* vol. 3, no. 5, pp. 409-416, 2018.
[http://dx.doi.org/10.1001/jamacardio.2018.0136] [PMID: 29562087]

[19]   H.Y.Y. Nyein, L.C. Tai, Q.P. Ngo, M. Chao, G.B. Zhang, W. Gao, M. Bariya, J. Bullock, H. Kim, H.M. Fahad, and A. Javey, "A wearable microfluidic sensing patch for dynamic sweat secretion analysis", *ACS Sens.,* vol. 3, no. 5, pp. 944-952, 2018.
[http://dx.doi.org/10.1021/acssensors.7b00961] [PMID: 29741360]

[20]   O. Parlak, S.T. Keene, A. Marais, V.F. Curto, and A. Salleo, "Molecularly selective nanoporous membrane-based wearable organic electrochemical device for noninvasive cortisol sensing", *Sci. Adv.,* vol. 4, no. 7, p. eaar2904, 2018.
[http://dx.doi.org/10.1126/sciadv.aar2904] [PMID: 30035216]

[21]   M.H. Iqbal, A. Aydin, O. Brunckhorst, P. Dasgupta, and K. Ahmed, "A review of wearable technology in medicine", *J. R. Soc. Med.,* vol. 109, no. 10, pp. 372-380, 2016.
[http://dx.doi.org/10.1177/0141076816663560] [PMID: 27729595]

[22]   J.M. Peake, G. Kerr, and J.P. Sullivan, "A critical review of consumer wearables, mobile applications, and equipment for providing biofeedback, monitoring stress, and sleep in physically active populations", *Front. Physiol.,* vol. 9, no. 743, p. 743, 2018.
[http://dx.doi.org/10.3389/fphys.2018.00743] [PMID: 30002629]

[23]   N. Constant, O. Douglas-Prawl, S. Johnson, and K. Mankodiya, "Pulse-Glasses: An unobtrusive, wearable HR monitor with Internet-of-Things functionality", *Proce. of the 2015 IEEE 12th Inter. Conf. on Wearable and Implantable Body Sensor Net. (BSN),* pp. 1-5 Cambridge, MA, USA 2015.
[http://dx.doi.org/10.1109/BSN.2015.7299350]

[24]   J.R. Sempionatto, L.C. Brazaca, L. García-Carmona, G. Bolat, A.S. Campbell, A. Martin, G. Tang, R. Shah, R.K. Mishra, J. Kim, V. Zucolotto, A. Escarpa, and J. Wang, "Eyeglasses-based tear biosensing system: Non-invasive detection of alcohol, vitamins and glucose", *Biosens. Bioelectron.,* vol. 137, pp. 161-170, 2019.
[http://dx.doi.org/10.1016/j.bios.2019.04.058] [PMID: 31096082]

[25]   T. Arakawa, and K. Mitsubayashi, Cavitas Sensors (Soft Contact Lens Type Biosensor, Mouth-Guard Type Sensor, etc.) for Daily Medicine.*Sensors for Everyday Life: Healthcare Settings.,* O.A. Postolache, S.C. Mukhopadhyay, K.P. Jayasundera, A.K. Swain, Eds., Springer International Publishing: Cham, Switzerlandpp. 45-65. 2017.
[http://dx.doi.org/10.1007/978-3-319-47319-2_3]

[26]   L. Van Langenhove, C. Hertleer, P. Westbroek, and J. Priniotakis, Textile sensors for healthcare. *Smart Textiles for Medicine and Healthcare,* L. Van Langenhove, Ed., Woodhead Publishing: Sawston, Cambridgepp. 106-122. 2007.

[27]   E. Ismar, S. Kurşun Bahadir, F. Kalaoglu, and V. Koncar, "Futuristic clothes: electronic textiles and wearable technologies", *Glob. Chall.,* vol. 4, no. 7, p. 1900092, 2020.
[http://dx.doi.org/10.1002/gch2.201900092]

[28]   X. Liu, and P.B. Lillehoj, "Embroidered electrochemical sensors for biomolecular detection", *Lab Chip,* vol. 16, no. 11, pp. 2093-2098, 2016.
[http://dx.doi.org/10.1039/C6LC00307A] [PMID: 27156700]

[29]   M. Stoppa, and A. Chiolerio, "Wearable electronics and smart textiles: a critical review", *Sensors (Basel),* vol. 14, no. 7, pp. 11957-11992, 2014.
[http://dx.doi.org/10.3390/s140711957] [PMID: 25004153]

[30]   R.K. Mishra, L.J. Hubble, A. Martín, R. Kumar, A. Barfidokht, J. Kim, M.M. Musameh, I.L. Kyratzis, and J. Wang, "Wearable flexible and stretchable glove biosensor for on-site detection of organophosphorus chemical threats", *ACS Sens.,* vol. 2, no. 4, pp. 553-561, 2017.
[http://dx.doi.org/10.1021/acssensors.7b00051] [PMID: 28723187]

[31]   R. Villar, T. Beltrame, and R.L. Hughson, "Validation of the Hexoskin wearable vest during lying, sitting, standing, and walking activities", *Appl. Physiol. Nutr. Metab.,* vol. 40, no. 10, pp. 1019-1024, 2015.
[http://dx.doi.org/10.1139/apnm-2015-0140] [PMID: 26360814]

[32]   S.K. Kundu, S. Kumagai, and M. Sasaki, "A wearable capacitive sensor for monitoring human respiratory rate", *J. Appl. Phys.,* vol. 52, pp. 04-05, 2013.
[http://dx.doi.org/10.7567/JJAP.52.04CL05]

[33]   M. Hyland, H. Hunter, J. Liu, E. Veety, and D. Vashaee, "Wearable thermoelectric generators for human body heat harvesting", *Appl. Energy,* vol. 182, pp. 518-524, 2016.
[http://dx.doi.org/10.1016/j.apenergy.2016.08.150]

[34]   Y.D. Lee, and W.Y. Chung, "Wireless sensor network based wearable smart shirt for ubiquitous health and activity monitoring", *Sens. Actuators B Chem.,* vol. 140, no. 2, pp. 390-395, 2009.
[http://dx.doi.org/10.1016/j.snb.2009.04.040]

[35]   M. Taj-Eldin, C. Ryan, B. O'Flynn, and P. Galvin, "A review of wearable solutions for physiological and emotional monitoring for use by people with Autism Spectrum Disorder and their caregivers", *Sensors (Basel),* vol. 18, no. 12, p. 4271, 2018.
[http://dx.doi.org/10.3390/s18124271] [PMID: 30518133]

[36]   V.Q. Le, T.H. Do, J.R.D. Retamal, P.W. Shao, Y.H. Lai, W.W. Wu, J.H. He, Y.L. Chueh, and Y.H. Chu, "Van der Waals heteroepitaxial AZO/NiO/AZO/muscovite (ANA/muscovite) transparent flexible memristor", *Nano Energy,* vol. 56, pp. 322-329, 2019.
[http://dx.doi.org/10.1016/j.nanoen.2018.10.042]

[37]   H.R. Lim, H.S. Kim, R. Qazi, Y.T. Kwon, J.W. Jeong, and W.H. Yeo, "Advanced Soft Materials, Sensor Integrations, and Applications of wearable flexible hybrid electronics in healthcare, energy, and environment", *Adv. Mater.,* vol. 32, no. 15, p. 1901924, 2020.
[http://dx.doi.org/10.1002/adma.201901924] [PMID: 31282063]

[38]   J. Kang, J.B-H. Tok, and Z. Bao, "Self-healing soft electronics", *Nat. Electron.,* vol. 2, no. 4, pp. 144-150, 2019.
[http://dx.doi.org/10.1038/s41928-019-0235-0]

[39]   A.J. Bandodkar, C.S. López, A.M. Vinu Mohan, L. Yin, R. Kumar, and J. Wang, "All-printed magnetically self-healing electrochemical devices", *Sci. Adv.,* vol. 2, no. 11, p. e1601465, 2016.
[http://dx.doi.org/10.1126/sciadv.1601465] [PMID: 27847875]

[40]   D. Son, and Z. Bao, "Nanomaterials in skin-inspired electronics: Toward soft and robust skin-like electronic nanosystems", *ACS Nano,* vol. 12, no. 12, pp. 11731-11739, 2018.
[http://dx.doi.org/10.1021/acsnano.8b07738] [PMID: 30460841]

[41]   D. Jiang, Y. Wang, B. Li, C. Sun, Z. Wu, H. Yan, L. Xing, S. Qi, Y. Li, H. Liu, W. Xie, X. Wang, T. Ding, and Z. Guo, "Flexible sandwich structural strain sensor based on silver nanowires decorated with self-healing substrate", *Macromol. Mater. Eng.,* vol. 304, no. 7, p. 1900074, 2019.
[http://dx.doi.org/10.1002/mame.201900074]

[42]   C. Pang, C. Lee, and K.Y. Suh, "Recent advances in flexible sensors for wearable and implantable devices", *J. Appl. Polym. Sci.,* vol. 130, no. 3, pp. 1429-1441, 2013.
[http://dx.doi.org/10.1002/app.39461]

[43]   F. Fu, J. Wang, H. Zeng, and J. Yu, "Functional conductive hydrogels for bioelectronics", *ACS Materials Letters,* vol. 2, no. 10, pp. 1287-1301, 2020.
[http://dx.doi.org/10.1021/acsmaterialslett.0c00309]

[44]   J. Chen, J. Zheng, Q. Gao, J. Zhang, J. Zhang, O. Omisore, L. Wang, and H. Li, "Polydimethylsiloxane (PDMS)-based flexible resistive strain sensors for wearable applications", *Appl. Sci. (Basel),* vol. 8, no. 3, p. 345, 2018.
[http://dx.doi.org/10.3390/app8030345]

[45]   L. Tang, S. Wu, J. Qu, L. Gong, and J. Tang, "A review of conductive hydrogel used in flexible strain sensor", *Materials (Basel),* vol. 13, no. 18, p. 3947, 2020.
[http://dx.doi.org/10.3390/ma13183947] [PMID: 32906652]

[46]   Y. Cao, Y.J. Tan, S. Li, W.W. Lee, H. Guo, Y. Cai, C. Wang, and B.C.K. Tee, "Self-healing electronic skins for aquatic environments", *Nat. Electron.,* vol. 2, no. 2, pp. 75-82, 2019.
[http://dx.doi.org/10.1038/s41928-019-0206-5]

[47]   V.K. Chaturvedi, N. Yadav, N.K. Rai, N.H.A. Ellah, R.A. Bohara, I.F. Rehan, N. Marraiki, G.E.S. Batiha, H.F. Hetta, and M.P. Singh, "Pleurotus sajor-caju-mediated synthesis of silver and gold nanoparticles active against colon cancer cell lines: A new era of herbonanoceutics", *Molecules,* vol. 25, no. 13, p. 3091, 2020.
[http://dx.doi.org/10.3390/molecules25133091] [PMID: 32645899]

[48]   C.M. Boutry, Y. Kaizawa, B.C. Schroeder, A. Chortos, A. Legrand, Z. Wang, J. Chang, P. Fox, and Z. Bao, "A stretchable and biodegradable strain and pressure sensor for orthopaedic application", *Nat. Electron.,* vol. 1, no. 5, pp. 314-321, 2018.
[http://dx.doi.org/10.1038/s41928-018-0071-7]

[49]   S.W. Hwang, H. Tao, D.H. Kim, H. Cheng, J.K. Song, E. Rill, M.A. Brenckle, B. Panilaitis, S.M. Won, Y.S. Kim, Y.M. Song, K.J. Yu, A. Ameen, R. Li, Y. Su, M. Yang, D.L. Kaplan, M.R. Zakin, M.J. Slepian, Y. Huang, F.G. Omenetto, and J.A. Rogers, "A physically transient form of silicon electronics", *Science,* vol. 337, no. 6102, pp. 1640-1644, 2012.
[http://dx.doi.org/10.1126/science.1226325] [PMID: 23019646]

[50]   D. Morris, S. Coyle, Y. Wu, K.T. Lau, G. Wallace, and D. Diamond, "Bio-sensing textile based patch with integrated optical detection system for sweat monitoring", *Sens. Actuators B Chem.,* vol. 139, no. 1, pp. 231-236, 2009.
[http://dx.doi.org/10.1016/j.snb.2009.02.032]

[51]   V.F. Curto, S. Coyle, R. Byrne, N. Angelov, D. Diamond, and F. Benito-Lopez, "Concept and development of an autonomous wearable micro-fluidic platform for real time pH sweat analysis", *Sens. Actuators B Chem.,* vol. 175, pp. 263-270, 2012.
[http://dx.doi.org/10.1016/j.snb.2012.02.010]

[52]   M. Caldara, C. Colleoni, E. Guido, V. Re, and G. Rosace, "Optical monitoring of sweat pH by a textile fabric wearable sensor based on covalently bonded litmus-3-glycidoxypropyltrimethoxysilane coating", *Sens. Actuators B Chem.,* vol. 222, pp. 213-220, 2016.
[http://dx.doi.org/10.1016/j.snb.2015.08.073]

[53]   X. Mu, X. Xin, C. Fan, X. Li, X. Tian, K.F. Xu, and Z. Zheng, "A paper-based skin patch for the diagnostic screening of cystic fibrosis", *Chem. Commun. (Camb.),* vol. 51, no. 29, pp. 6365-6368, 2015.
[http://dx.doi.org/10.1039/C5CC00717H] [PMID: 25761978]

[54]   B. Schazmann, D. Morris, C. Slater, S. Beirne, C. Fay, R. Reuveny, N. Moyna, and D. Diamond, "A wearable electrochemical sensor for the real-time measurement of sweat sodium concentration", *Anal. Methods,* vol. 2, no. 4, pp. 342-348, 2010.

[http://dx.doi.org/10.1039/b9ay00184k]

[55] G. Matzeu, C. Fay, A. Vaillant, S. Coyle, and D. Diamond, "A wearable device for monitoring sweat rates *via* image analysis", *IEEE Trans. Biomed. Eng.,* vol. 63, no. 8, pp. 1672-1680, 2016.
[http://dx.doi.org/10.1109/TBME.2015.2477676] [PMID: 26394409]

[56] A. Martín, J. Kim, J.F. Kurniawan, J.R. Sempionatto, J.R. Moreto, G. Tang, A.S. Campbell, A. Shin, M.Y. Lee, X. Liu, and J. Wang, "Epidermal microfluidic electrochemical detection system: Enhanced sweat sampling and metabolite detection", *ACS Sens.,* vol. 2, no. 12, pp. 1860-1868, 2017.
[http://dx.doi.org/10.1021/acssensors.7b00729] [PMID: 29152973]

[57] T. Pfaffe, J. Cooper-White, P. Beyerlein, K. Kostner, and C. Punyadeera, "Diagnostic potential of saliva: current state and future applications", *Clin. Chem.,* vol. 57, no. 5, pp. 675-687, 2011.
[http://dx.doi.org/10.1373/clinchem.2010.153767] [PMID: 21383043]

[58] R.S.P. Malon, S. Sadir, M. Balakrishnan, and E.P. Córcoles, "Saliva-based biosensors: noninvasive monitoring tool for clinical diagnostics", *BioMed Res. Int.,* vol. 2014, pp. 1-20, 2014.
[http://dx.doi.org/10.1155/2014/962903] [PMID: 25276835]

[59] M. Falk, V. Andoralov, Z. Blum, J. Sotres, D.B. Suyatin, T. Ruzgas, T. Arnebrant, and S. Shleev, "Biofuel cell as a power source for electronic contact lenses", *Biosens. Bioelectron.,* vol. 37, no. 1, pp. 38-45, 2012.
[http://dx.doi.org/10.1016/j.bios.2012.04.030] [PMID: 22621980]

[60] J. Heikenfeld, "Technological leap for sweat sensing", *Nature,* vol. 529, no. 7587, pp. 475-476, 2016.
[http://dx.doi.org/10.1038/529475a] [PMID: 26819041]

[61] Z. Sonner, E. Wilder, J. Heikenfeld, G. Kasting, F. Beyette, D. Swaile, F. Sherman, J. Joyce, J. Hagen, N. Kelley-Loughnane, and R. Naik, "The microfluidics of the eccrine sweat gland, including biomarker partitioning, transport, and biosensing implications", *Biomicrofluidics,* vol. 9, no. 3, p. 031301, 2015.
[http://dx.doi.org/10.1063/1.4921039] [PMID: 26045728]

[62] J.R. Windmiller, A.J. Bandodkar, G. Valdés-Ramírez, S. Parkhomovsky, A.G. Martinez, and J. Wang, "Electrochemical sensing based on printable temporary transfer tattoos", *Chem. Commun. (Camb.),* vol. 48, no. 54, pp. 6794-6796, 2012.
[http://dx.doi.org/10.1039/c2cc32839a] [PMID: 22669136]

[63] D.P. Rose, M.E. Ratterman, D.K. Griffin, L. Hou, N. Kelley-Loughnane, R.R. Naik, J.A. Hagen, I. Papautsky, and J.C. Heikenfeld, "Adhesive RFID sensor patch for monitoring of sweat electrolytes", *IEEE Trans. Biomed. Eng.,* vol. 62, no. 6, pp. 1457-1465, 2015.
[http://dx.doi.org/10.1109/TBME.2014.2369991] [PMID: 25398174]

[64] J. Ferreira Gonzalez, "Textile-enabled bioimpedance instrumentation for personalised health monitoring applications", *Licentiate Thesis,* KTH Royal Institute of Technology: Stockholm, Sweden, 2013.

[65] A. Poscia, M. Mascini, D. Moscone, M. Luzzana, G. Caramenti, P. Cremonesi, F. Valgimigli, C. Bongiovanni, and M. Varalli, "A microdialysis technique for continuous subcutaneous glucose monitoring in diabetic patients (part 1)", *Biosens. Bioelectron.,* vol. 18, no. 7, pp. 891-898, 2003.
[http://dx.doi.org/10.1016/S0956-5663(02)00216-6] [PMID: 12713912]

[66] S.A.N. Gowers, V.F. Curto, C.A. Seneci, C. Wang, S. Anastasova, P. Vadgama, G.Z. Yang, and M.G. Boutelle, "3D printed microfluidic device with integrated biosensors for online analysis of subcutaneous human microdialysate", *Anal. Chem.,* vol. 87, no. 15, pp. 7763-7770, 2015.
[http://dx.doi.org/10.1021/acs.analchem.5b01353] [PMID: 26070023]

CHAPTER 3

3D Bioprinting of Advanced Bioinks for Tissue Regeneration and Biosensor Development

Alma Tamunonengiofori Banigo[1,*] and **Chinedu Chamberlin Obasi**[2]

[1] *Department of Developmental Bio Engineering, Faculty of Science and Technology and TechMed Centre, University of Twente. Drienerlolaan 5, 7522 NB Enschede, The Netherlands*

[2] *Department of Pharmaceutical Biotechnology, Institute of Pharmacy, Martin Luther University, Halle-Wittenberg, Germany*

Abstract: Three-dimensional (3D) bioprinting technology has become a unique system for tissue regeneration and biosensor development by controlled deposition of bioinks to produce complex constructs. Different bioprinters including laser-assisted and extrusion-based have been introduced and used to produce constructs with high resolution, cell viability and shape fidelity for tissue development. In addition, microfluidic technology, organ-on-a-chip and electrospinning technology are used to produce biosensing products to diagnose and monitor living systems. One of the most critical materials used for bioprinting is bioink. Several bioinks of an advanced level and different compositions have been developed too. Here, we briefly highlighted the characteristics, advantages, and disadvantages of some bioprinters and advanced bioinks that have been developed recently. We also stated some tissue engineering applications with the use of 3D bioprinting. Lastly, we mentioned a few key areas for main focus in the future.

Keywords: Applications, Advanced boinks, Bioprinting, Hydrogel, Tissue regeneration, Tissue engineering.

INTRODUCTION

Worn-out or damaged living tissues occur as a common phenomenon in humans. The ability to repair or regenerate these tissues that are damaged from diseases is termed tissue regeneration [1]. The conventional techniques employed for tissue regeneration depend on the transplantation of tissues. Unfortunately, transplantation cannot be initiated in this case without the presence of a donor, which could be scarce and may bring about the risk of graft rejection due to an immune response.

* **Corresponding author Alma Tamunonengiofori Banigo:** Department of Developmental Bio Engineering, Faculty of Science and Technology and TechMed Centre, University of Twente. Drienerlolaan 5, 7522 NB Enschede, The Netherlands; E-mail: a.tamunonengioforibanigo@utwente.nl

Vivek K. Chaturvedi, Dawesh P. Yadav and Mohan P. Singh (Eds.)

Tissue engineering (TE) is a branch of biomedical engineering (BME) that focuses on tackling these challenges of tissue regeneration [2]. Specifically, additive manufacturing (AM) is a technique in TE that has the principles of some fields, including biology and material science [3, 4], that can perform the following functions:-

1) To produce novel tissues for restoring damaged ones;

2) To mimic the native complexity of the cell micro-environment;

3) To facilitate cellular activities, including differentiation, proliferation, and tissue regeneration [5].

The conventional methods lack the 3D cell distribution [6], take time, are less efficient [5], and many more [5, 7]. AM enables the construction of complex tissues based on a top-down approach by using precise and accurate computer-generated 3D tissue models [8].

3D bioprinting is a branch of AM [5] that has been expanding rapidly and currently, focuses on producing biomimetic tissue-like constructs. This technology utilizes some imaging techniques, such as computed tomography (CT) and magnetic resonance imaging (MRI), to make a computer-aided design (CAD) [9]. The design is used for layer-by-layer (bottom-up approach) deposition of biomaterials, viable cells, and other supporting materials (including growth factors, and nucleic acids) into accurate and precise geometries onto a substrate [10], and they are positioned precisely with functional components to fabricate tissue-like 3D structures [11] or constructs with unique structure and function [9, 11]. Based on the presence of cells, tissues, and other living components, observations such as viability, the biocompatibility of the present materials, cell sensitivity to the printing techniques, perfusion, toxicity and many more have to be considered [5, 12]. These bioprinted tissues with enhanced properties and intercellular communication could help in modeling the human *in vivo* physiology [5]. Hence, the outcome can be used in pre-clinical trials as animal models lack the ability to detect human pathophysiological responses [13]. The models from the expansion of 3D bioprinting technology can be used for different applications, including:-

- *In situ* surgical cartilage repair [14]
- *In vitro* disease modeling [9]
- Drug screening, disease mechanism research, and pre-clinical studies in tissues of interest [15 - 24]

- *In situ* wound dressing [25]
- Biosensor development [26 - 28]

For the latter aspect 'biosensor development', 3D bioprinting technologies are being employed to develop biosensing platforms including impedimetric-based sensing platforms to monitor both conditions of living cells (proliferation and activity in either suspensions or adherent cultures) and bioink (composition and quality) during continuous micro extrusion [27, 29]. In comparison with traditional 3D printing to produce cell-free scaffolds, 3D bioprinting needs several technical requirements and approaches to fabricate 3D constructs with both biological and mechanical properties for functional tissue restoration and biosensor development. The 3D bioprinting process, some advantages of 3D bioprinting over traditional 3D printing, and several approaches to 3D bioprinting have been outlined in Figs. (**1a, b and c**), respectively.

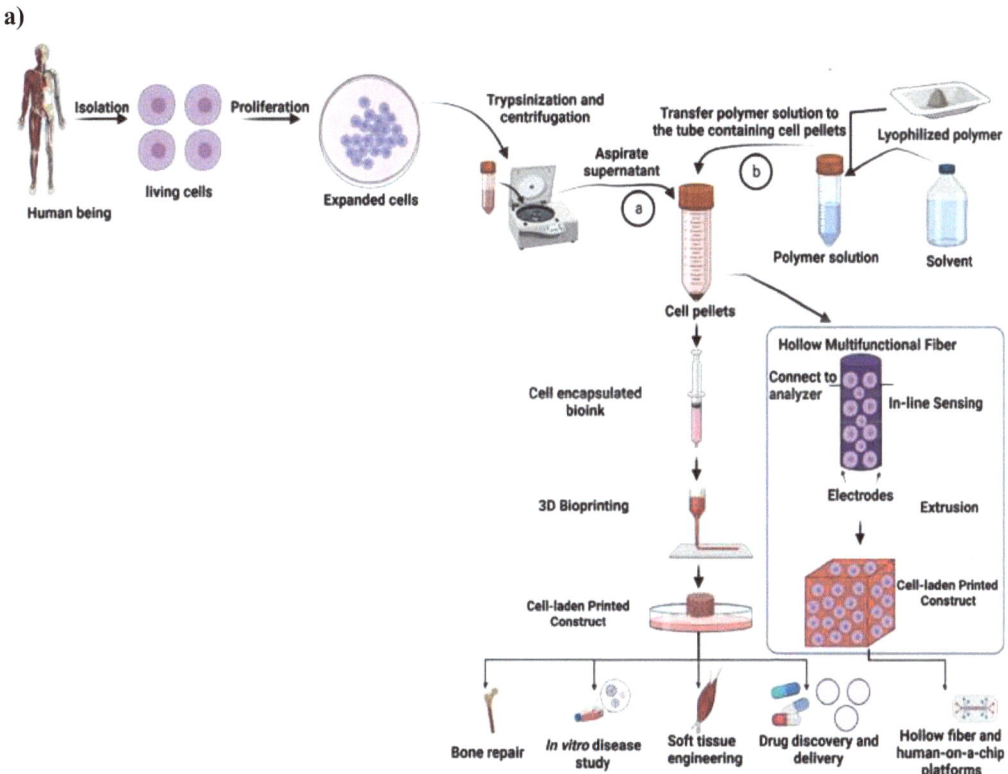

(Fig. 1) contd.....

b)

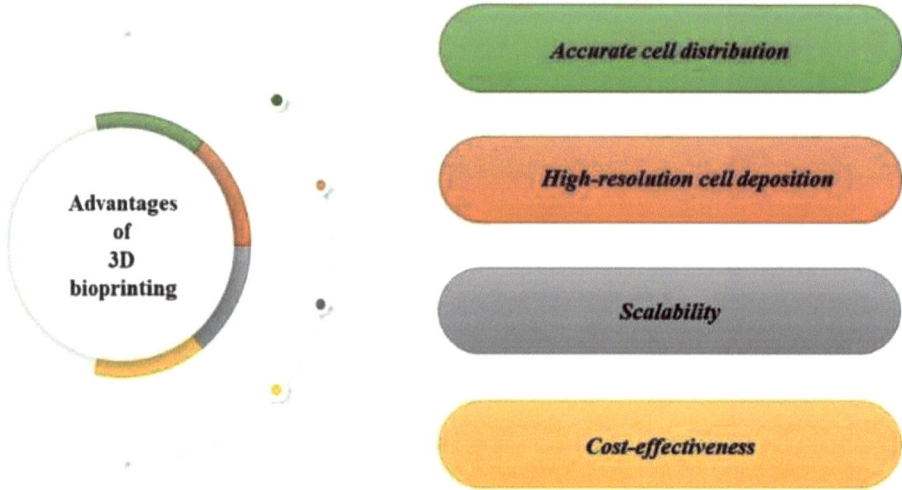

c)

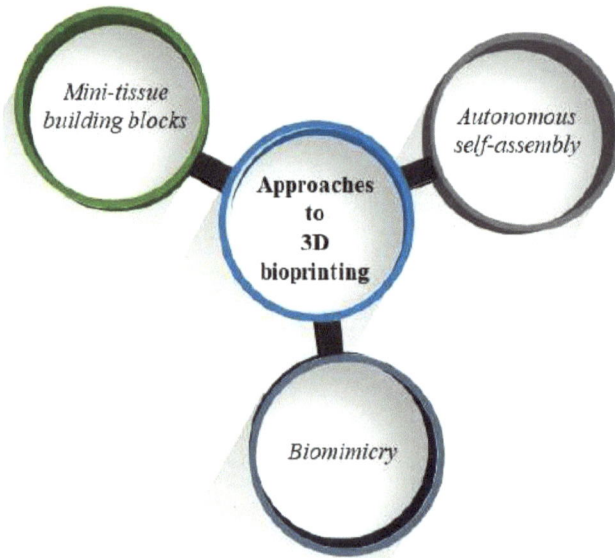

Fig. (1). *3D bioprinting technology.* **a)** Typical examples of the processes involved (created with BioRender.com) [27, 40] **b)** Advantages of 3D bioprinting over traditional 3D printing **c)** Different approaches considered (created with SlideModel.com).

Despite the above-mentioned advantages, the full potential of the technology has not been reached due to some limiting factors. These factors may include the development of suitable bioinks, the improvement of the printing process for faster printing speeds and optimal scalability, and better resolution of the printed constructs with enhanced biological and mechanical properties. The development of bioink is very crucial in this review, and it is basically defined as a flowing material consisting of biomaterials, growth factors, cells, and other materials [30, 31]. Various tissue types, including skin, heart, liver, and bone, can be produced by this technology, along with producing microfluidic models of organs-on-a-chip in the near future [32]. The realization of a functional tissue construct depends on vascularization of the tissue, gas and nutrient exchange, biocompatibility, biodegradability of the material (used as a substrate), shape fidelity, and preservation of the printed tissue functionality [33].

In terms of biosensor development, a typical challenge in the area of process monitoring is the utilization of cells from animals, or sometimes human cells, as processed materials to serve as living materials from the processing area. Also, there is a dire need for sensing platforms to monitor both cell quality measures and bioink composition [27].

Based on these reasons, bioinks comprising of natural and synthetic polymers, including alginate, gelatin, hyaluronic acid, polyethylene glycol (PEG), and many more, possessing both biological and controlled physio-chemical properties can be modified to fit the structure and formation of the extracellular matrix (ECM) [34, 35]. These bioinks are termed 'advanced bioinks' with the intention to enhance the functionality of the printed scaffolds far away from the traditional paradigm of the 'biofabrication window' as shown in Table (**1**) and Fig. (**2a**) [36]. Major characteristics and categories of these advanced bioinks are highlighted in Figs. (**2b**) and (**2c**), and more information can be obtained from [36].

With a proper study of the working principles of TE, the bioprinting technologies can be categorized into 4 groups, namely inkjet-based, extrusion-based, stereolithography, and laser-assisted [37]. To print 3D biomimetic constructs, suitable bioinks should be carefully developed. To successfully develop the bioinks, biomaterials should possess the necessary requirements such as biocompatibility, biostability, biodegradability, bioprintability and structural integrity after printing [37 - 39]. To achieve this developmental goal, advanced bioinks have been and are currently being formulated with utmost consideration of their cellular viability and mechanical properties; especially, the combination of two or more biomaterials has shown to be potential advanced bioink sources.

Research-based on the combination of two or more bioinks has shown to be a vital tool for tissue regeneration. Cells of different types can be encapsulated into the bioinks to yield stunning results in the repair of tissues such as bone, blood vessels, cartilage, and many more. In as much as many researchers have reported studies associated with 3D bioprinting, this review aims to highlight the characteristics, advantages, and disadvantages of some bioprinters and advanced bioinks. Here, we present the combination of 2 or more bioinks for proper use in TE applications. We also stated some tissue engineering applications with the use of 3D bioprinting. Future perspectives are also considered briefly.

Table 1. Differences between the traditional and advanced bioinks (created with SlideModel.com) [36].

Traditional

Advanced

- Ability to house encapsulated cells.
- Compromise between suitability for fabrication.

- High print fidelity.
- Shear-thinning.
- High mechanical strength.
- High cytocompatibility.
- Ability to modulate cellular functions.

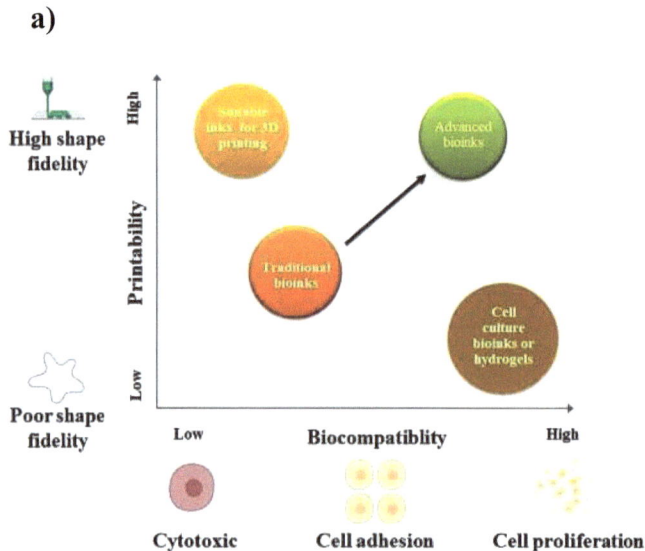

a)

(Fig. 2) contd.....

b)

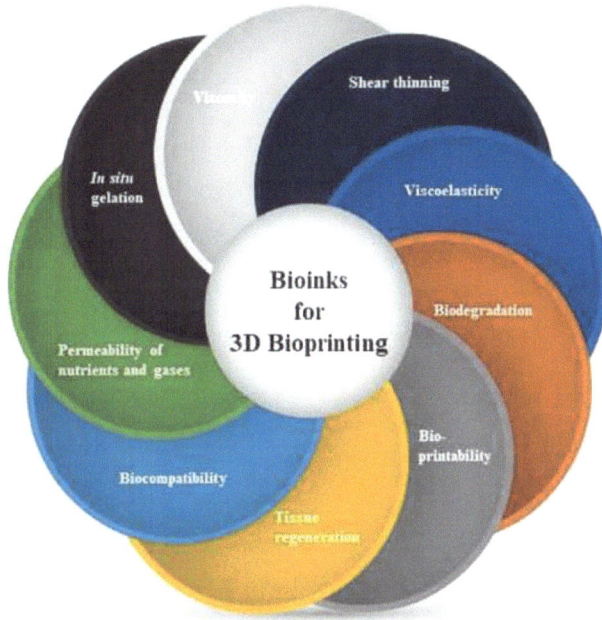

c)

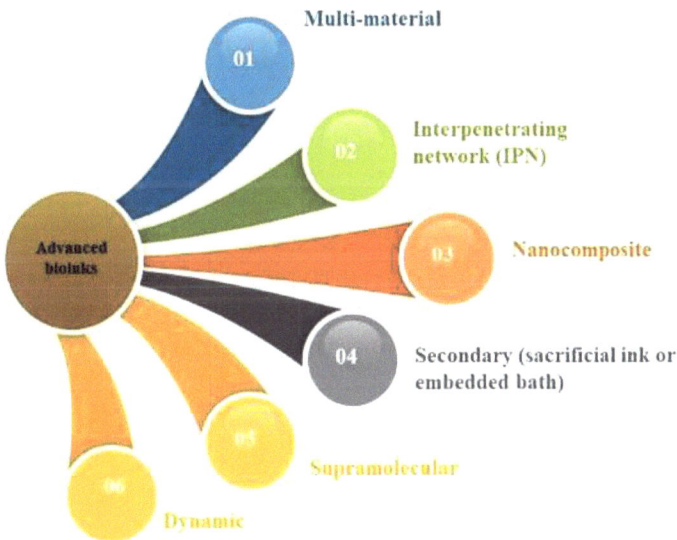

Fig. (2). *Advancedbioinks for 3D bioprinting* **a)** The biofabrication window for the design of advanced bioinks requires a balance between printability and biocompatibility **b)** Unique properties of an advanced bioink **c)** Categories of advanced bioinks into six major groups (created with SlideModel.com) [36].

3D Bioprinters

Tissue Regeneration

As stated earlier, the 3D Bioprinters are listed in 4 groups considering the working principles. Here, we mentioned more than four types of bioprinters with a focus on the use of advanced bioinks. They include inkjet-based, extrusion-based, stereolithography, laser-assisted, fused deposition modeling, acoustic, microvalve, and needle array bioprinters [31, 37]. Fig. (3) depicts 5 types of bioprinters. We briefly stated the overview of each working principle and the important characteristics of 4 major types. The choice of the right bioprinter should be dependent on many factors, including the structural properties of the tissue or organ to be replaced.

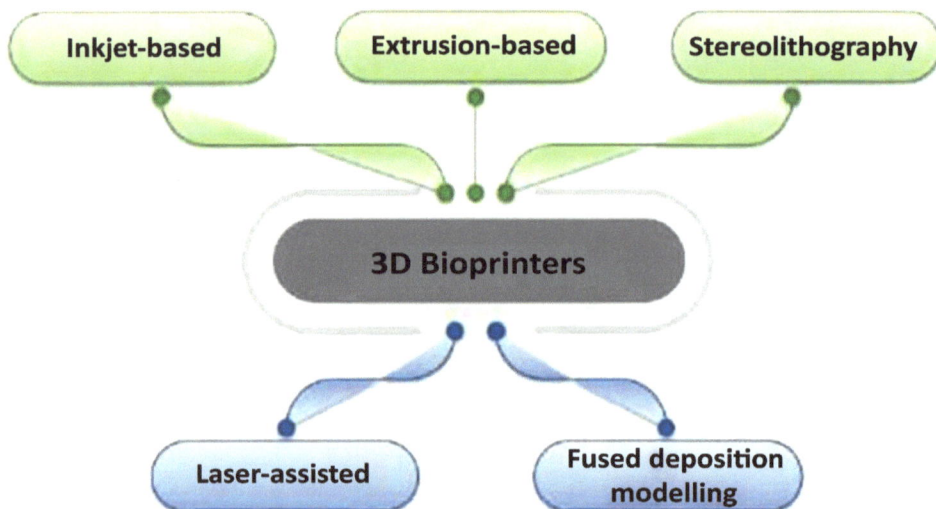

Fig. (3). Various types of 3D Bioprinters (created with SlideModel.com).

Inkjet-Based Bioprinters

The inkjet-based bioprinter was brought to the limelight in 1988 by Klebe, of which a Hewlett-Packard (HP) inkjet bioprinter with a hydrogel solution was used to print by him [41]. Inkjet-based 3D bioprinting involves the use of a bioink which comprises a low viscous ink and living cells that can be unloaded on a bio paper, petri dish, or a polymer construct. It is also a non-contact printing method that occurs in a digitally controlled way. This bioprinting technology can be achieved in two simple forms, either continuous inkjet printing or a drop-on demand manner. More information on both forms can be found in research articles [5, 37]. This inkjet-based bioprinting technology uses materials with a viscosity of 3.5 x 12 mPa.s and different crosslinking strategies.

The printing modules in the inkjet bioprinter were focused on depositing either the biomaterials or cells as a droplet unit through different dispensing units (*via* heating reservoirs or piezoelectric actuators [5, 37]. The important characteristics of the types of inkjet-based bioprinters, advantages, and disadvantages are summarized in (Table **2**).

Table 2. *Inkjet-based bioprinters.* a) **Types of the 3D bioprinter and the advanced bioprinting techniques b) Advantages and disadvantages of using the inkjet-based bioprinting techniques [12, 31, 42-46].**

a)

S/No.	Types of the Inkjet-based bioprinter		Advanced bioprinting techniques
	Thermal	Piezoelectric	
1	Uses a heating element (HE) that superheats the bioink by increasing the temperature up to 300 °C.	Uses a voltage to introduce a rapid shape change in the piezoelectric material.	Modified inkjet-based techniques termed multi-jets have been introduced to produce complex tissues and organs prototypes with proper cell type's arrangement as well as other tissue parts.
2	Gasification can occur as bubbles are generated and are forcefully printed as droplets on the printing bed.	Pressure pulses are introduced in the fluid which pushes out the cell-laden or non-cell-laden droplets through the attached nozzle	
3	------------	More varieties of inks can be used without coagulation.	

b)

Advantages

- Uses high print speed and low cost.

- Causes high resolution and employs simple processing technique.

- Droplet densities and sizes can be increased to bring in concentration gradients of cells, materials or growth factors in the printed constructs.

- Acoustic radiation can generate and control uniform droplet size and ejection directionality.

- Layers of cells can be printed in a sacrificial mold.

- Quick production with highly repeatable structure.

Disadvantages

- Requires narrow range biomaterial viscosities.

- Uses limited cell densities and mechanical strength.

- Cell damage or cell lysis may occur while printing from both types of bioprinter due to the acoustic frequencies used.

- Non-uniform droplet sizes may be produced.
- Clogging of the nozzle may occur.

- The inkjet printing dimensions are limited by the diameter of the inkjet needle.

Extrusion-Based (EB) Bioprinters

The extrusion-based printers deposit (bio)inks layer by layer either by the pneumatic system or mechanical tools (piston or screw). The first EB printer was mentioned in 2002 [47]. The printers, along with bioinks can be used to produce large and complex cell-encapsulated or cell-free biomimetic structures consisting of two or more materials or cell spheroids [48]. This EB technology uses a wide range of materials with a viscosity between 30 and 6 x 10^7mPa.s [49] and also applies different crosslinking mechanisms. The printers are quite affordable and are divided into three main types, namely the pneumatic, piston, and screw-driven. More information about the different characteristics of the types of extrusion-based bioprinters, advantages, and disadvantages is summarized in (Table **3**).

Some companies, including Cellink, Allevi, and Advanced solutions, developed extrusion-based bioprinters for the production of living tissue constructs [31, 50]. Amongst all the available bioprinting companies, Cellink company was introduced in 2016 with the production of the first cost-affordable 3D bioprinter model INKREDIBLE available for commercial purchase. This bioprinter had ultraviolet (UV) crosslinking at 365 nm LED (405 nm, optional) but without thermal regulation for both print beds and extruders. Afterward, improved versions such as INKREDIBLE +, BIO X, and BIO X6 were produced [31].

Researchers have used Cellink's INKREDIBLE + 3D bioprinter and bioinks to print autologous nasal cartilage grafts [51, 52] and in several scientific research articles. Currently, our team at the Department of Developmental Bio Engineering at the University of Twente has successfully used the same Cellink's INKREDIBLE + 3D bioprinter, nearly instantaneous enzymatic crosslinking mechanism, and tyramine-modified high molecular weight (HMW) hyaluronic acid (HA-TA) bioink for coaxial bioprinting of cell-laden filaments in one step for the first time. This research will be published soon. Based on the results obtained, further steps are being explored, with the use of the 3D bioprinter, coaxial nozzles, and advanced bioinks to print more filaments and hollow tubes with higher cell viability and better mechanical properties, which is the first step to successful printing of living fibers including peripheral nerve fibers. Also, tubular-structured organs with improved vascularization will be bioprinted too.

The next bioprinter produced by Cellink, the Bio X 3D Bioprinter model, has improved and upgraded built-in temperature-controlled print bed (4-60 ^{0}C), the dualH14 HEPA filters, and UV-C lamp 275nm. This tool was used by researchers to produce 3D bioprinted scaffolds for diabetic wound-healing applications [53] and many more.

Table 3. *Extrusion-based bioprinters.* **a)** Types of the 3D bioprinter and the advanced bioprinting techniques. **b)** Advantages and disadvantages of using the extrusion-based bioprinting techniques [52, 54-56].

a)

S/No.	Types of the Extrusion-based bioprinter			Advanced bioprinting techniques
	Pneumatic System	**Mechanical tools**		
		Piston	**Screw**	
1	Uses pressurized air flow with valve-free or valve-dependent configurations. The valve-based system can control either pressure or pulse frequency yielding a more exact application. The valve-free systems are simple but low viscous bioinks can leak out of the nozzle with little or no control.	Uses an axial piston and can control the flow of the bioink through the nozzle.	Uses screws as well as permits better positional control and works better with highly viscous bioinks.	Several advanced EBB techniques such as multi-material, coaxial, triaxial, and hybrid bioprinting techniques are used to produce simple and complex structures for tissue regeneration.
2	Simple and highly viable in material deposition and works better for highly viscous materials.	Not suitable for highly viscous bioinks.		

b)

Advantages	**Disadvantages**
· Several biomaterial types can be used with a wide range of viscosities.	· The nozzle in the printer can cause relatively low resolution and poor cell viability due to the presence of shear stress from the systems (mechanical force or pressure).
· High cell seeding density and different crosslinking mechanisms can be applied.	· Highly viscous bioink induces higher shear stress that can lead to cell death.
· Higher printing speed may be employed to facilitate scalability.	· Very low viscous bioinks yield poor print fidelity.
· 3D tissue constructs are produced by a layer-by-layer deposition method.	· Clogging of the nozzle can occur.
· Shorter fabrication time is achieved with this bioprinter better than the laser-assisted and inkjet-based methods.	

Laser-Assisted Bioprinters (LAB)

The laser-assisted bioprinters with the use of optical cell trapping were first announced by D. Odde in the late 1990s [57]. For LAB, the pulsed laser beam is required to deposit bioinks (cells and biomaterials) in a petri dish or on a substrate. LAB shows more promising outcomes than laser-guided direct writing meant for the patterning of cells and tissue engineering. This technique may also manage single-cell deposits at a rate of kHz [58]. A popular LAB comprises more than two parts: a pulsed laser source, a receiving substrate, and a target coated with the potential printing material [5]. The characteristics of the laser-assisted bioprinter, its advantages, and its disadvantages were summarized in (Table **4**).

More advantages and disadvantages of inkjet-based, extrusion-based, and laser-assisted bioprinting technologies have been written in some research and review articles [53, 59 - 62].

Table 4. *Laser-assisted bioprinter.* **a) Description of the 3D bioprinter and the advanced bioprinting techniques. b) Advantages and disadvantages of using the laser-assisted bioprinting techniques [5, 57 & 58].**

a)

S/No.	Laser-assisted bioprinter	Advanced bioprinting techniques
1	Uses a laser to give light to the minute area of the donor ribbon layer, and a high-pressure bubble is produced. The bubble moves the layer of the bioink as droplets are made so that the bioink could be unloaded onto the substrate.	There can be more opportunities in the medical robotics area, where *in vivo* laser-assisted bioprinting and printing of both biomaterials and cells could take place. LAB method with no metallic interlayer could be fabricated to prevent the cytotoxicity of its residues.
2	There is a low chance of contamination because the bioinks and the dispenser do not meet each other.	Combining LAB with other laser-assisted techniques such as polymerization and machining could help subcellular resolution.
3	The LAB mechanism depends on several parameters.	

b)

Advantages	Disadvantages
· Works without a nozzle.	· Uses a long fabrication time for some biomaterials.
· Can print viscous biomaterials.	· Has poor mechanical properties.
· Gives good resolution, speed, and accuracy.	· The laser can induce cell damage.
· No nozzle clogging due to the nozzle-free printing system.	· Expensive printing modules.
· Deposition of little volume (smaller than a picolitre is possible).	· Difficult to use.

Stereolithography Bioprinters

The stereolithographic type of bioprinting concentrates on the design height because it makes the design in a layered manner in the presence of light on a light-based heat curable bioink including gelatin methacrylamide (GelMA) or polyethylene glycol diacrylate (PEGDA) [63], in a plane by plane manner [64]. The stereolithography 3D printing machine called the stereo lithograph apparatus (SLA) has four different components, namely a tank for the photopolymer, a perforated material placed at the bottom of the tank, an ultraviolet (UV) and a computer device controlling the material, and the laser. The premier stereolithography was first announced by C.W. Hull in the late 1980s [37].

The two main groups of stereolithographic printing are single-photon and multiphoton groups, which will be further divided into subgroups. More information about the printer, its subgroups, advantages, and advanced stereolithographic printing techniques is highlighted in (Table **5**).

Table 5. *Stereolithography bioprinter.* a) Description of the 3D bioprinter, its subgroups and the advanced bioprinting techniques. b) Advantages and disadvantages of using the stereolithography bioprinting techniques [5, 37].

a)

S/No.	Stereolithography bioprinter	Advanced bioprinting techniques
1	Uses light to crosslink the bioinks in the reservoir via a layer-by-layer technique. This technique can be coordinated with clinical imaging techniques including MRI to enhance the diagnostic techniques, quality, and design of the medical devices and the relevant success of complex surgeries.	Low force stereolithography (LFS) technology is the future phase in stereolithography bioprinting, meeting the demands of the current market for scalable, reliable, and industrial-quality 3D printing.
2	There are two groups of this technique such as single-photon and multiphoton. The subgroup of single-photon includes visible radiation, IR stereolithography, stereo-thermal lithography, and conventional stereolithography systems. Two-photon laser scanning photolithography can be used to produce 3D liver tissue structures.	This LFS technique decreases the forces on printing areas with the use of a flexible tank and linear illumination to give better surface quality and accuracy of the print.
3	Stereolithography is used for producing biocompatible scaffold in TE where resins help to avoid inflammatory responses during implantation.	

b)

Advantages	Disadvantages
• This bioprinter creates prototypes because it produces highly accurate, durable objects so fast and affordable.	• This method is limited to using light-based bioinks.
• The equipment can produce objects with different shapes which may be difficult to make with the traditional prototyping methods.	• The reservoir can also be filled with photopolymers, which comprises of material waste and expensive experimentation.
• This bioprinter uses less printing time to produce constructs with a high degree of accuracy.	• The post-process takes time and its highly complicated.
• This technique yields high cell viability and promotes the growth of BrCa that led to a good model for post-metastatic breast cancer progression investigation in a bone.	

Other Bioprinters

There are other existing bioprinters, such as acoustic and microvalve bioprinters. The acoustic and microvalve are examples of droplet-based bioprinters. The former bioprinter deposits droplets as a force produced with the use of acoustic waves [65]. This former bioprinter does not allow the living cells in the bioinks to come in contact with heat or high pressure that induces cell death [66] if compared with the inkjet and extrusion-based bioprinter. The latter (microvalve) bioprinter deposits droplets with the use of an electromechanical microvalve comprising a valve coil and plunger [66]. In short, the valve coil produces a magnetic field, which pushes the plunger upwards. The bioinks present in the barrel are deposited through the unblocked barrel by the use of pneumatic pressure. The droplets produced from the latter printer are bigger than the ones produced from the inkjet-based with the same nozzle size and the resolution is decreased [66]. Many bioprinters, including needle array bioprinters, have been produced for different purposes [37].

The individual bioprinting technique has its own unique characteristics. In order to produce complex biomimetic structures, the application of more than one printing technique (hybrid technique) can be made. Combinations of inkjet-based, laser-assisted, or extrusion-based printers with electrospinning [67 - 69], as well as integrated inkjet-based and extrusion-based bioprinters [70], were explored by researchers. Hybrid techniques have been used to improve the mechanical strength and biological performance of the cell-laden construct or printed structure [37] and engineer 3D skin models [70] better than traditional 3D bioprinting. In addition to that, the printing techniques use more difficult and complex production stages as well as software and hardware systems which make the intended researchers face challenges while using these printing techniques [37].

BIOSENSOR DEVELOPMENT

There are two groups of 3D printed sensors, namely engineering-based and medicine-based groups. The engineering-based group produces mechanical, temperature, particle, and tactile sensors. Then, the medicine-based group focuses on biomolecules-based, microbial, bionic, and cell-based sensors [71]. 3D printed technologies have been introduced to produce biosensing products for biomedical applications. The micro extrusion 3D bioprinter, microfluidic technology, organ-on-chip technology, and electrospinning technology, among others, are used to produce biosensing devices or products (including fibers) [72] for medical diagnosis or monitoring of some systems, including micro physiological systems [73]. The fibers can be used in the biomedical field for various applications related to chemical, medical, and physical sensors that could monitor the above-mentioned purposes, including stress, moisture, and temperature. Some of the various sensors produced by organ-on-chip technology are electrochemical (target particular cell materials for monitoring cellular performance), pH, and oxygen sensors. More information can be obtained from [40] and other articles not mentioned here. Despite the advantages of the biosensing devices and products mentioned or not, many challenges (such as miniaturization, the improvement of their biocompatibility, long-term stability of ligands, low limits of detection, flexibility, stretchability, and durability) are yet to be tackled for the successful addition of biosensors into organ-on-chip technology [74 - 76]. Some typical examples of bioprinted tissue [72], fiber [27], and bionic sensors [77] can be shown in Figs. (**4**, **5** and **6**).

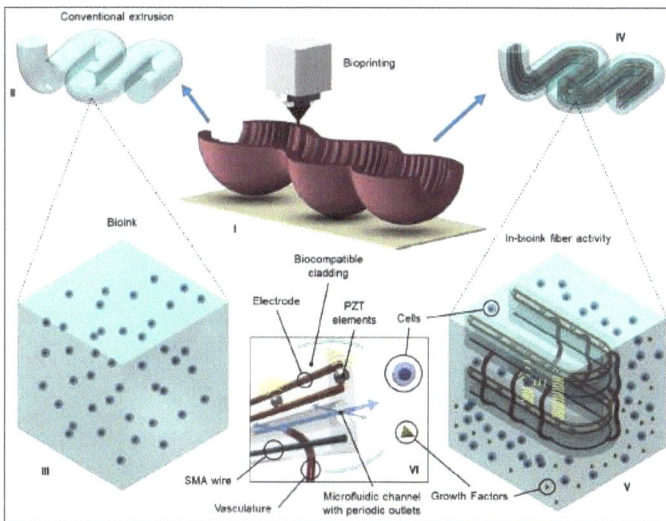

Fig. (4). *Biological interfacing* (**I**) 3D bioprinting of a tissue (**II**) Standard micro extrusion of bioink (**III**) Conventional bioink with cells suspended in the hydrogel (**IV**) New coaxial micro extrusion of bio interfacing fiber coated in bioink (**V**) A clear view of fiber where bio interfacing occurs: epithelial cells and vascular epithelial growth factors are excreted from different microchannels and yield in cellular self-assembled vasculature between two orifices; piezoelectric elements measure surrounding cell density by ultrasound and shape memory alloy wires provide peristaltic motion in the tissue (**VI**) Visualization of bio interfacing fiber and its components [72].

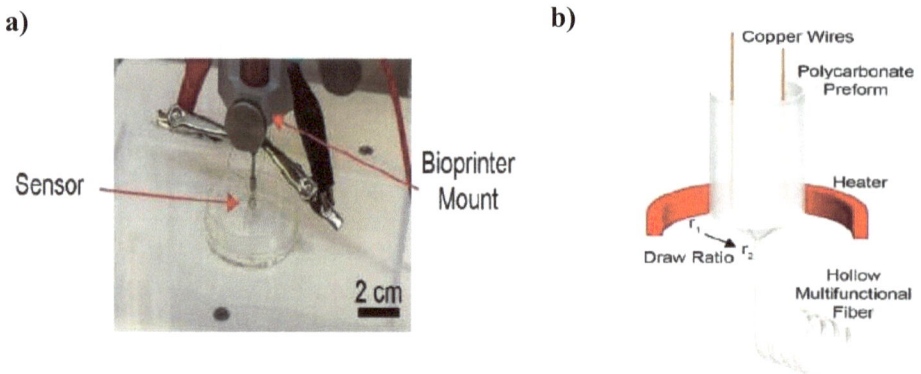

Fig. (5). *Hollow multifunctional fibers* **a)** A photographic image showing 3D bioprinting through the hollow multifunctional fiber impedimetric sensor. **b)** A schematic representation of the manufacturing process to develop electrode-functionalized hollow multifunctional fibers. A preform tube with embedded copper wires was drawn, and the drawdown ratio controlled the fiber diameter. Cross sections of drawn fibers with inner diameters of 300 [27].

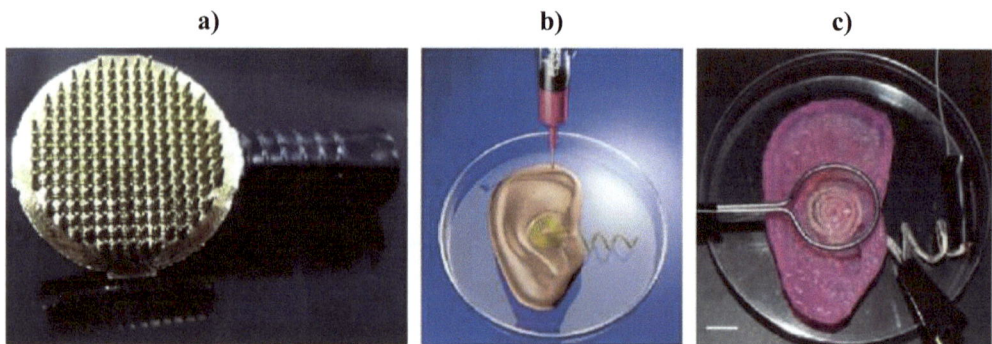

Fig. (6). *A bionic ear* **a)** The 3D printed electrode **b)** A computer diagram of a bionic ear **c)** Co-printing of a conductor to produce a bionic ear [71, 77].

ADVANCED BIOINKS

Multi-Material Bioinks

Single-material bioinks usually produce bioprinted constructs for TE applications with limited biological or physicochemical properties and other issues [78]. A typical example is our HMW HA-TA bioink used for coaxial bioprinting of core filaments. The bioink had several properties, including biodegradability and shear thinning properties but inhibited good cell proliferation and migration. To solve this problem and many more from other research with the use of a single bioink, combining bioinks with unique properties, or adding fillers and additives with different properties (including nanoparticles) can fabricate 3D bioprinted composite. This 3D Bioprinted composite will have the following properties:-

1) Optimal mechanical properties

2) High performance

3) High fidelity

4) Good biocompatibility

5) Enhanced biomimicking of tissue constructs [78]

For instance, combining HMW HA-TA with some existing materials such as arginyl-glycyl-aspartic acid [RGD] (present in gelatin) or many more will improve the bioactivity of our HMW HA-TA bioink. Some scientific authors have already combined materials for the successful bioprinting of tissue constructs. Li *et al.* [79] successfully applied alginate and methyl cellulose (Alg/MC) blend hydrogel for bioprinting. The utilization of the blended material yielded good extrudability, adhesion between printed layers, high thixotropic property and stackability [79]. Post-crosslinking further improved the adhesion strength between the printed layers. Also, the introduction of a chelating agent to eradicate the superficial calcium ions led to better printability, higher shape fidelity and stackability [79].

Duan *et al.* [80] also combined methacrylated hyaluronan (HAMA) and gelatin methacrylate (GelMA) to print 3D trileaflet heart valves. GelMA enhanced the adhesion characteristics of the cell and HAMA increased the viscosity of the bioink. Many more multi-material bioinks have been used to also produce big and complex structures by heterogeneous layer-by-layer printing where one of the bioinks may be used as a temporary (sacrificial) support that will be washed away after printing to prevent the construct from collapsing [78].

Secondary (Sacrificial Bioink or Embedding Bath)

Here, a sacrificial ink such as Pluronic F127 can be used to give temporary support [81, 82] while printing with either a coaxial, triaxial, or multi-head bioprinting technique. Coaxial bioprinting involves the controlled deposition of bioinks simultaneously from a coaxial nozzle. This technique can be used to print bioinks and crosslinking agents side by side to enhance the resolution of the printed fibers or hollow structures, instead of depending on using a crosslinking bath. A typical example is the production of a bio-blood-vessel (BBV) by the use of a coaxial cell printing technique *via* the extrusion of a Pluronic ink (CPF-127) (inside) and a composite bioink composed of vascular-tissue derived ECM (VdECM) and alginate (outside) [82]. This composite bioink used yielded some benefits, including differentiation, cell proliferation, neovascularization and production of tubular BBVs [78].

Kim *et al.* [83] also used the coaxial printing technique to produce tubular constructs and studied their structural integrity and shape fidelity. Other researchers, including our team, have explored the use of coaxial bioprinting to print core filaments, hollow structures [82 - 84], and fibers [85] for different applications. Our team is currently focusing on using advanced bioinks (HA and gelatin, dextran, or bio gel if possible) in the core region, with the sacrificial bioink in the sheath region, to print core filaments. The core filaments are expected to have high printability, shape fidelity, and bioactivity after proper optimization and study.

Alternately, researchers use hydrogel-embedded baths for bioink printability. Hinton *et al.* [86] developed a composite bioink (made of collagen, Matrigel, fibrinogen, and hyaluronic acid) to produce complex structures by embedding the printed structure within a secondary "sacrificial" hydrogels (gelatin slurry). After printing the structure, the gelatin support bath was removed by heating the bath to a physiological temperature. Models of complex structures, such as a human right coronary arterial tree and an explanted embryonic chick heart, can be printed with high structural fidelity [36].

Some researchers also combined the use of an embedded bath with another bioprinting technique, such as coaxial bioprinting, to print 3D tissue constructs. Hong *et al.* [84] utilized Gel-PEG-TA composite bioink with an enzymatic crosslinking mechanism and coaxial bioprinting to produce vascular constructs in the presence of a fibrin bath. They combined both coaxial bioprinting and the use of a support bath to produce a complex structure.

Interpenetrating Networks (IPNs) Bioinks

IPNs comprise individual polymer networks that are physically entangled with one another [36]. These bioinks use several chemistries for crosslinking to allow each polymer network to be involved in intra-crosslinking [87]. Limited inter-crosslinking also takes place but in a different form. The use of IPNs improves:

• Both toughness and fracture strength are based on the individual component networks of each polymer constituent [87, 88].

Double network (DN) is a branch of IPNs that is massively involved in producing DN hydrogels but is quite slow for 3D bioprinting [36]. Then, ionic-covalent entanglement (ICE) hydrogels were produced by physical and chemical crosslinking for 3D printing applications [87, 89]. For instance, Bakarich *et al.*

[90] utilized ICE hydrogels consisting of acrylamide and alginate for 3D bioprinting. The properties (such as the printed shape and stiffness) of the composite hydrogel were enhanced. In the second research study, Hong *et al.* [84] produced elastomeric ICE hydrogels composed of poly(ethylene glycol) diacrylate (PEGDA) and alginate. Fracture strength improved and mechanical stress was sustained during the process. Interestingly, IPNs can be further utilized with nanomaterials to produce structures with high print fidelity [36].

Nanocomposite Bioinks

Nanocomposite bioinks can be used to print 3D tissue constructs for TE applications. The bioinks comprising polymers and little amount of nanoparticles show the relevant difference in physical and chemical properties, which include bioactivity,controlled drug release,photo responsiveness,resistance to degradation within physiological conditions,electrical conductivity, shear-thinning properties and stiffness [91-94].

Nanocomposite bioinks consisting of hydroxyapatite nanoparticles (nHAp) material and poly(ethylene glycol) (PEG) showed high compressive modulus and osteogenic potential with the use of human mesenchymal stem cells [36]. Another nanocomposite bioink (nano fibrillated cellulose and alginate) with good shear-thinning properties was used to produce soft tissue constructs with high structural fidelity at 25 °C [95]. Then, GelMA and nano silicates were used to promote osteogenic differentiation and facilitate the development of mineralized ECM. These studies and others not mentioned show the possibility and importance of using nanocomposite bioinks to improve the 3D printing process as well as contribute to the bioactivity of encapsulated cells [36].

Supramolecular

Some constructs or hydrogels are required to be mechanically tough and should withstand mechanical deformation [36]. In a situation where they break due to repeated stress, they lose mechanical integrity. In order to solve this issue, supramolecular bioinks should be developed. They are made up of short repeating units with functional groups that can interact non-covalently with other functional units to produce large, polymer-like entanglements [36].

Highley *et al.* [96] showed a simple technique to produce shear-thinning and mechanically resilient hydrogels for 3D bioprinting applications with hyaluronic acid (HA)-based supramolecular hydrogel. The quick increase in viscosity after printing provides high structural fidelity and integrity [36].

Li *et al.* [97] also developed supramolecular polypeptide–DNA hydrogel for quick *in situ* 3D bioprinting by developing a composite bioink (polypeptide–DNA conjugate and complementary DNA linker). The rigidity of DNA polymers allows for the printing of constructs with high structural integrity [36]. The constructs had high cytocompatibility and biodegradability [36].

Dynamic Bioinks

Dynamic bioinks for tissue engineering, regenerative medicine, and 3D bioprinting should be a natural evolution. These bioinks require nonlinear viscoelastic properties meant for 3D bioprinting [98].

Current studies focus on supramolecular and dynamic covalent chemistry (DCvC) to create a new path to producing advanced bioinks. DCvC is a process that reversibly gives covalent bonds in fixed states. The physiologically important DCvC has been applied in different areas, including drug delivery and biosensing [98].

With the advancement of the molecular structure of the dynamic reactive constituents, impacts can be made on the equilibrium and rate constants as well as the mechanical properties [99]. The molecular dials in DCvC can influence the degree of thixotropy to give optimal bioprinting as you vary the bulk stiffness and relaxation dynamics for optimal cell behaviour [98]. With extrusion-based bioprinting, researchers used composite materials, including two side-chain modified hyaluronic acid macromers functionalized with aldehydes (~21 kDa) or hydrazides (74 kDa) to generate DCvC hydrazone crosslinks, which were used during 3D bioprinting [88]. The crosslinker resulted in self-healing, shear-thinning, and injectable materials.

Despite the advantages (both mentioned or not), there is still more to be done to use the dynamic bioinks for successful 3D bioprinting of living constructs. They are as follows:-

• Association cooperativity
• Quantification of the bioink printability focused on mechanical characterization
• Important cellular timescales
• Mechanical tuning and harnessing different ranges of dynamicity simultaneously
• stiffness and tunability should be known
• optimize bioprintability
• processability [98]

Above all, dynamic bioinks and other advanced bioinks have a bright future in the 3D bioprinting of simple and complex tissues. (Table **6**) depicts the description the above-mentioned advanced bioinks and (Table **7**) highlights advantages, and disadvantages of these advanced bioinks.

Table 6. The description of the advanced bioinks for 3D bioprinting [36, 84, 85, 98].

S/No.	Advanced bioink types	Bioprinting simple and complex constructs
		Description
1	Multi-material	These are widely investigated bioinks that overcome the mysteries of one-component hydrogels. For instance, high molecular weight (HMW) hyaluronic acid (HA) or alginate has been used as a single-component hydrogel in TE because of their unique properties. HMW HA possesses high viscosity, good biodegradability, shear thinning, and anti-inflammatory properties and can be enzymatically crosslinked using horseradish peroxidase and hydrogen peroxide (as we have shown in our recent research). More so, HMW HA hinders cell migration and proliferation. On the other hand, the alginate biomaterial is biocompatible and could be ionically crosslinked in the presence of calcium ions to produce hydrogels or constructs but this material is bioinert and can cause cell death. Combining two materials such as the HMW HA and gelatin or alginate and gelatin will improve the printing performance, biological and rheological properties.
2	Secondary	Sacrificial ink or embedded bath can be used to provide temporary support during the printing of the constructs. Both conditions can be used together with multi-material to print complex structures.
3	Interpenetrating network	Interpenetrating networks (IPNs) are composite hydrogels with individual networks entangled within each other physically. The primary and secondary networks comprise an elastic polymer and stiff polymer respectively. Ionic and covalent (ICE) hydrogels can be produced to improve different properties.
4	Nanocomposite	Nanocomposite hydrogels give an easy tactic to join different functionalities in the 3D printed constructs by adding nanoparticles with good characteristics. Nanocomposite hydrogels are used for tissue engineering applications and 3D bioprinting.
5	Supramolecular	Supramolecular bioinks have self-healing, shear-thinning, and stress-relaxation properties, which support the bioink performance during printing and afterward. Supramolecular bioinks are developed to solve the problem of mechanical integrity loss. In the presence of high stress, the bonds are reversibly broken to dissipate energy. The reversibility of these bonds gives shear-thinning properties that help the 3D bioprinting process.
6	Dynamic	Dynamic covalent chemistry (DCvC) reactions include disulfide exchange, boronate diels-alder reactions in reverse order, ester and aldimine formation.

Table 7. Advantages, and disadvantages of using the advanced bioinks for 3D bioprinting [36, 84, 85, 98].

S/No.	Advanced bioink types	Bioprinting simple and complex constructs	
		Advantages	**Disadvantages**
1	Multi-material	A unique approach to enhance printability, structural fidelity, viscosity, cell adhesion, and biocompatibility without any form of compromise. The long-term mechanical properties can be kept sub-optimal. This technique can yield enhanced collagen and glycosaminoglycan which can show ECM remodelling. Maintenance of the fibroblastic phenotype of cells can occur. Complex and big structures can be produced A combination of different properties in an experiment with high strength and functionality is possible.	Interfacial performance between the different materials may affect the structural reliability of the composite material.
2	Secondary	This type induces differentiation, cell proliferation and Neovascularisation. The use of a secondary bioink can assist in fabricating tubular BBVs with high structural fidelity.	For coaxial bioprinting, the number of materials and cell types is limited when compared with multi-head or triaxial bioprinting.
3	Interpenetrating network	These advanced bioinks improve toughness and fracture strength. Increase stiffness and failure stress. ICE can produce mechanically tough 3D bioprinted constructs for different applications.	DN is quite slow for 3D bioprinting.
4	Nanocomposite	Nanocomposite bioinks improve compressive modulus, mechanical strength, print fidelity, and shear-thinning properties and promote osteogenic differentiation of hMSCs. The use of IPNs and nanomaterials can increase print fidelity.	
5	Supramolecular	This advanced bioink can give high cytocompatibility, biodegradability, structural fidelity, and integrity.	The bonds lack mechanical strength for long-term stability.
6	Dynamic	The dynamic bioink allows tailorability, gives better printing performance, and produces biomimetic microenvironments for tissue maturation.	

APPLICATIONS

There is a rapidly growing request for tissue transplantation due to a shortage of donors [37]. Several tissues have been produced *via* 3D bioprinting, namely blood vessels, cartilage, bone, skin, liver, and neural tissues. To enhance the printability, bioactivity, and biomimicry of the bioinks or to increase the mechanical properties of the constructs, advanced bioinks have been and are still being developed with two or more bioinks. Here, we briefly highlighted the TE applications focused on 3D bioprinting in Fig. (**7**) [36, 37, 81, 83, 84, 90, 98, 100 - 102] and an example of the biosensors for the human-on-a-chip in Fig. (**8**).

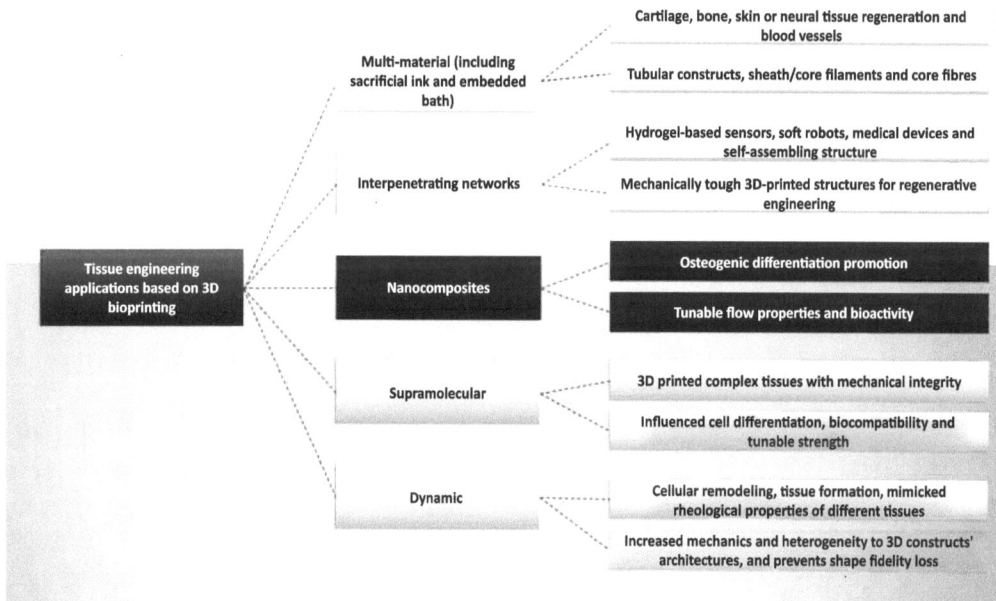

Fig. (7). Tissue engineering applications based on 3D bioprinting of advanced bioinks (created with SlideModel.com).

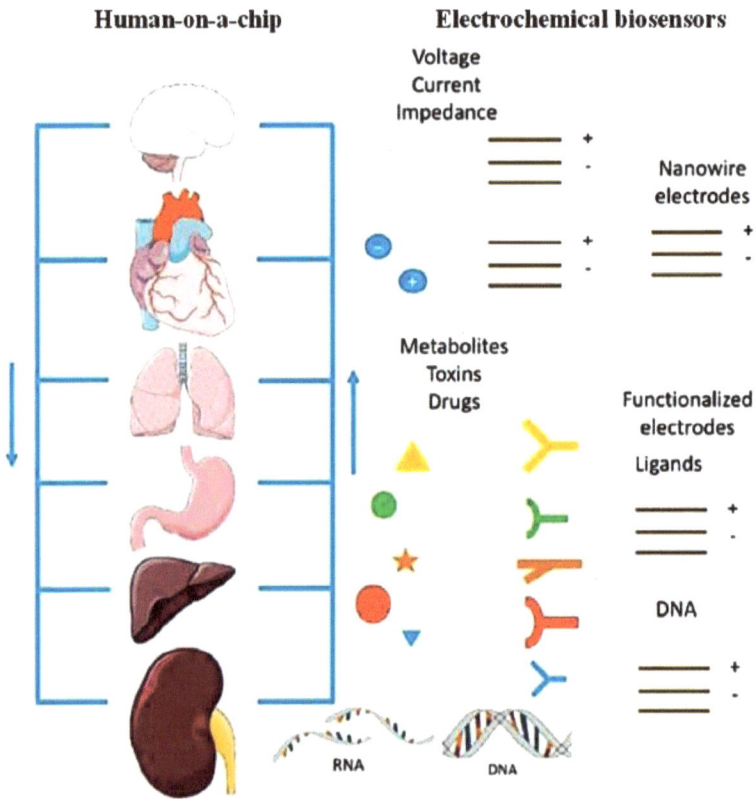

Fig. (8). A typical example of the biosensors for the human-on-a-chip [40].

CONCLUSION AND FUTURE PERSPECTIVES

The choice of advanced bioinks is an important condition for successful 3D bioprinting. The advanced bioinks should have several properties, including mechanical, rheological, and biological properties. Several bioinks have been produced and used for different purposes, and efforts to develop better bioinks should be made to solve existing limitations. Researchers should consider focusing on bioprintability and biofunctionality [37] as well as the chosen tissue type. In conclusion, advanced bioinks for 3D bioprinting may give better environments than traditional bioinks.

More so, human-on-a-chip platforms will be produced in the future. The platform with added biosensors will tailor towards the advancement of organ-on-a-chip platforms, giving physiological metabolism parameters of the organ model, as seen in Fig. (**8**). With this, novel substantial experiments will be carried out to enhance our understanding of the evolution of particular pathologies and the way they affect the entire system [40].

LIST OF ABBREVIATIONS

ALG	Alginate
AM	Additive manufacturing
BBV	Bio-blood-vessel
BME	Biomedical engineering
BrCa	Breast cancer gene
CAD	Computer-aided design
CPF-127	Chitosan-g-pluronic F-127
CT	Computed tomography
3D	Three-dimensional
^{0}C	Degree Celsius
DCvC	Dynamic covalent chemistry
DN	Double network
DNA	Deoxyribonucleic acid
EB	Extrusion-based
EBB	Extrusion-based bioprinter
ECM	Extracellular matrix
GelMA	Gelatin methacrylamide
Gel-PEG-TA	Gelatin-Polyethylene glycol-tyramine
hMSCs	Human mesenchymal stem cells
HA	Hyaluronic acid
HAMA	Methacrylated hyaluronan
HA-TA	Hyaluronic acid-tyramine
HE	Heating element
HEPA	High efficiency particulate air
HMW	High molecular weight
HP	Hewlett-Packard
IR	Infrared
ICE	Ionic-covalent entanglement
IPNs	Interpenetrating networks
kDA	Kilo Daltons
kHz	One thousand hertz
LAB	Laser-assisted bioprinter
LED	Light-emitting diode
LFS	Low force stereolithography

MC	Methylcellulose
MRI	Magnetic resonance imaging
nHAp	Hydroxyapatite nanoparticles
NM	Nanometer
PEG	Polyethylene glycol
PEGDA	Polyethylene glycol diacrylate
PH	Potential of hydrogen
RGD	Arginyl-glycyl-aspartic acid
SLA	Stereo lithograph apparatus
TE	Tissue Engineering
UV	Ultraviolet
UV-C	Germicidal Ultraviolet-C
VdECM	Vascular-tissue derived extracellular matrix

ACKNOWLEDGEMENT

Declared none.

REFERENCES

[1]　P. Boisseau, and B. Loubaton, "Nanomedicine, nanotechnology in medicine", *C. R. Phys.,* vol. 12, no. 7, pp. 620-636, 2011.
[http://dx.doi.org/10.1016/j.crhy.2011.06.001]

[2]　S. Bose, M. Roy, and A. Bandyopadhyay, "Recent advances in bone tissue engineering scaffolds", *Trends Biotechnol.,* vol. 30, no. 10, pp. 546-554, 2012.
[http://dx.doi.org/10.1016/j.tibtech.2012.07.005] [PMID: 22939815]

[3]　S. Bose, and A. Bandyopadhyay, Additive manufacturing: the future of manufacturing in a flat world. *Additive manufacturing.* 2nd ed. CRC Press: Boca Raton, FL. 451-461, 2019.
[http://dx.doi.org/10.1201/9780429466236-15]

[4]　A. Bandyopadhyay, Y. Zhang, and S. Bose, "Recent developments in metal additive manufacturing", *Curr. Opin. Chem. Eng.,* vol. 28, pp. 96-104, 2020.
[http://dx.doi.org/10.1016/j.coche.2020.03.001] [PMID: 32432024]

[5]　S. Agarwal, A. Saha, V.K. Balla, A. Pal, A. Barui, and S. Bodhak, "Current developments in 3D bioprinting for tissue and organ regeneration-A review", *Front. Mech. Eng.,* vol. 6, no. 589171, pp. 1-22, 2020.
[http://dx.doi.org/10.3389/fmech.2020.589171]

[6]　S. Bose, S. Vahabzadeh, and A. Bandyopadhyay, "Bone tissue engineering using 3D printing", *Mater. Today,* vol. 16, no. 12, pp. 496-504, 2013.
[http://dx.doi.org/10.1016/j.mattod.2013.11.017]

[7]　P. Singh, and D.J. Williams, "Cell therapies: realizing the potential of this new dimension to medical therapeutics", *J. Tissue Eng. Regen. Med.,* vol. 2, no. 6, pp. 307-319, 2008.
[http://dx.doi.org/10.1002/term.108] [PMID: 18618613]

[8]　F.P.W. Melchels, M.A.N. Domingos, T.J. Klein, J. Malda, P.J. Bartolo, and D.W. Hutmacher, "Additive manufacturing of tissues and organs", *Prog. Polym. Sci.,* vol. 37, no. 8, pp. 1079-1104,

2012.
[http://dx.doi.org/10.1016/j.progpolymsci.2011.11.007]

[9] Y.J. Shin, R.T. Shafranek, J.H. Tsui, J. Walcott, and A. Nelson, "3D bioprinting of mechanically tuned bioinks derived from cardiac decellularized extracellular matrix", *Acta Biomater.,* vol. 119, no. 2021, pp. 75-88, 2020.
[http://dx.doi.org/10.1016/j.actbio.2020.11.006] [PMID: 33166713]

[10] J. Xiongfa, Z. Hao, Z. Liming, and X. Jun, "Recent advances in 3D bioprinting for the regeneration of functional cartilage", *Regen. Med.,* vol. 13, no. 1, pp. 73-87, 2018.
[http://dx.doi.org/10.2217/rme-2017-0106] [PMID: 29350587]

[11] C. Mandrycky, Z. Wang, K. Kim, and D.H. Kim, "3D bioprinting for engineering complex tissues", *Biotechnol. Adv.,* vol. 34, no. 4, pp. 422-434, 2016.
[http://dx.doi.org/10.1016/j.biotechadv.2015.12.011] [PMID: 26724184]

[12] S.V. Murphy, and A. Atala, "3D bioprinting of tissues and organs", *Nat. Biotechnol.,* vol. 32, no. 8, pp. 773-785, 2014.
[http://dx.doi.org/10.1038/nbt.2958] [PMID: 25093879]

[13] N. Shanks, R. Greek, and J. Greek, "Are animal models predictive for humans?", *Philos. Ethics Humanit. Med.,* vol. 4, no. 1, p. 2, 2009.
[http://dx.doi.org/10.1186/1747-5341-4-2] [PMID: 19146696]

[14] S. Duchi, C. Onofrillo, C.D. O'Çonnell, R. Blanchard, C. Augustine, A.F. Quigley, R.M.I. Kapsa, P. Pivonka, G. Wallace, C. Di Bella, and P.F.M. Choong, "Handheld Co-Axial Bioprinting: Application to *in situ* surgical cartilage repair", *Sci. Rep.,* vol. 7, no. 1, p. 5837, 2017.
[http://dx.doi.org/10.1038/s41598-017-05699-x] [PMID: 28724980]

[15] H.W. Kang, S.J. Lee, I.K. Ko, C. Kengla, J.J. Yoo, and A. Atala, "A 3D bioprinting system to produce human-scale tissue constructs with structural integrity", *Nat. Biotechnol.,* vol. 34, no. 3, pp. 312-319, 2016.
[http://dx.doi.org/10.1038/nbt.3413] [PMID: 26878319]

[16] D. Bejleri, and M.E. Davis, "Decellularized extracellular matrix materials for cardiac repair and regeneration", *Adv. Healthc. Mater.,* vol. 8, no. 5, p. 1801217, 2019.
[http://dx.doi.org/10.1002/adhm.201801217] [PMID: 30714354]

[17] Y.J. Choi, Y.J. Jun, D.Y. Kim, H.G. Yi, S.H. Chae, J. Kang, J. Lee, G. Gao, J.S. Kong, J. Jang, W.K. Chung, J.W. Rhie, and D.W. Cho, "A 3D cell printed muscle construct with tissue-derived bioink for the treatment of volumetric muscle loss", *Biomaterials,* vol. 206, pp. 160-169, 2019.
[http://dx.doi.org/10.1016/j.biomaterials.2019.03.036] [PMID: 30939408]

[18] A. Lee, A.R. Hudson, D.J. Shiwarski, J.W. Tashman, T.J. Hinton, S. Yerneni, J.M. Bliley, P.G. Campbell, and A.W. Feinberg, "3D bioprinting of collagen to rebuild components of the human heart", *Science,* vol. 365, no. 6452, pp. 482-487, 2019.
[http://dx.doi.org/10.1126/science.aav9051] [PMID: 31371612]

[19] L.L. Wang, C.B. Highley, Y.C. Yeh, J.H. Galarraga, S. Uman, and J.A. Burdick, "Three‐dimensional extrusion bioprinting of single and double–network hydrogels containing dynamic covalent crosslinks", *J. Biomed. Mater. Res. A,* vol. 106, no. 4, pp. 865-875, 2018.
[http://dx.doi.org/10.1002/jbm.a.36323] [PMID: 29314616]

[20] K.H. Song, C.B. Highley, A. Rouff, and J.A. Burdick, "Complex 3D-printed microchannels within cell-degradable hydrogels", *Adv. Funct. Mater.,* vol. 28, no. 31, p. 1801331, 2018.
[http://dx.doi.org/10.1002/adfm.201801331]

[21] L. Ouyang, C.B. Highley, C.B. Rodell, W. Sun, and J.A. Burdick, "3D printing of shear-thinning hyaluronic acid hydrogels with secondary cross-linking", *ACS Biomater. Sci. Eng.,* vol. 2, no. 10, pp. 1743-1751, 2016.
[http://dx.doi.org/10.1021/acsbiomaterials.6b00158] [PMID: 33440472]

[22]   A. Farzin, A.K. Miri, F. Sharifi, N. Faramarzi, A. Jaberi, A. Mostafavi, R. Solorzano, Y.S. Zhang, N. Annabi, A. Khademhosseini, and A. Tamayol, "3D printed sugar-based stents facilitating vascular anastomosis", *Adv. Healthc. Mater.,* vol. 7, no. 24, p. 1800702, 2018.
[http://dx.doi.org/10.1002/adhm.201800702] [PMID: 30375196]

[23]   K. Zhu, S.R. Shin, T. van Kempen, Y.C. Li, V. Ponraj, A. Nasajpour, S. Mandla, N. Hu, X. Liu, J. Leijten, Y.D. Lin, M.A. Hussain, Y.S. Zhang, A. Tamayol, and A. Khademhosseini, "Gold nanocomposite bioink for printing 3D cardiac constructs", *Adv. Funct. Mater.,* vol. 27, no. 12, p. 1605352, 2017.
[http://dx.doi.org/10.1002/adfm.201605352] [PMID: 30319321]

[24]   K.A. Homan, D.B. Kolesky, M.A. Skylar-Scott, J. Herrmann, H. Obuobi, A. Moisan, and J.A. Lewis, "Bioprinting of 3D convoluted renal proximal tubules on perfusable chips", *Sci. Rep.,* vol. 6, no. 1, p. 34845, 2016.
[http://dx.doi.org/10.1038/srep34845] [PMID: 27725720]

[25]   H. Li, F. Cheng, D.P. Orgill, J. Yao, and Y.S. Zhang, "Handheld bioprinting strategies for *in situ* wound dressing", *Essays Biochem.,* vol. 65, no. 3, pp. 533-543, 2021.
[http://dx.doi.org/10.1042/EBC20200098] [PMID: 34028545]

[26]   A.T. Banigo, C.A. Nnadiekwe, and E.M. Beasi, "Recent Advances in biosensing in tissue engineering and regenerative medicine", In: *Biosignal Processing.* IntechOpen: London, United Kingdom, 2022. Working Title
[http://dx.doi.org/10.5772/intechopen.104922]

[27]   A.P. Haring, S. Jiang, C. Barron, E.G. Thompson, H. Sontheimer, J.Q. He, X. Jia, and B.N. Johnson, "3D bioprinting using hollow multifunctional fiber impedimetric sensors", *Biofabrication,* vol. 12, no. 3, p. 035026, 2020.
[http://dx.doi.org/10.1088/1758-5090/ab94d0] [PMID: 32434163]

[28]   N. Wang, K. Burugapalli, S. Wijesuriya, M.Y. Far, W. Song, F. Moussy, Y. Zheng, Y. Ma, Z. Wu, and K. Li, "Electrospun polyurethane-core and gelatin-shell coaxial fibre coatings for miniature implantable biosensors", *Biofabrication,* vol. 6, no. 1, p. 015002, 2013.
[http://dx.doi.org/10.1088/1758-5082/6/1/015002] [PMID: 24346001]

[29]   W. Gamal, H. Wu, I. Underwood, J. Jia, S. Smith, and P. Bagnaninchi, "Impedance-based cellular assays for regenerative medicine", *Phil. Trans. R. Soc. B,* vol. 373, no. 1750, p. 20170226, 2018.
[http://dx.doi.org/10.1098/rstb.2017.0226]

[30]   I.T. Ozbolat, "Scaffold-based or scaffold-free bioprinting: competing or complementing approaches?", *J. Nanotechnol. Eng. Med.,* vol. 6, no. 2, p. 024701, 2015.
[http://dx.doi.org/10.1115/1.4030414]

[31]   A. Tong, Q.L. Pham, P. Abatemarco, A. Mathew, D. Gupta, S. Iyer, and R. Voronov, "Review of low-cost 3D bioprinters: state of the market and observed future trends", *SLAS Technol.,* vol. 26, no. 4, pp. 333-366, 2021.
[http://dx.doi.org/10.1177/24726303211020297] [PMID: 34137286]

[32]   F. Guillemot, B. Guillotin, A. Fontaine, M. Ali, S. Catros, V. Kériquel, J.C. Fricain, M. Rémy, R. Bareille, and J. Amédée-Vilamitjana, "Laser-assisted bioprinting to deal with tissue complexity in regenerative medicine", *MRS Bull.,* vol. 36, no. 12, pp. 1015-1019, 2011.
[http://dx.doi.org/10.1557/mrs.2011.272]

[33]   C. Xu, W. Chai, Y. Huang, and R.R. Markwald, "Scaffold-free inkjet printing of three-dimensional zigzag cellular tubes", *Biotechnol. Bioeng.,* vol. 109, no. 12, pp. 3152-3160, 2012.
[http://dx.doi.org/10.1002/bit.24591] [PMID: 22767299]

[34]   A. Tevlek, and H.M. Aydin, "Poly (glycerol-sebacate) elastomer: a mini review", *Ortho Surg. Open Access (OSOA. 000507).,* vol. 1, no. 2, pp. 1-4, 2017.

[35]   S. Bodhak, S. Bose, and A. Bandyopadhyay, "Electrically polarized HAp-coated Ti: *In vitro* bone

cell–material interactions", *Acta Biomater.,* vol. 6, no. 2, pp. 641-651, 2010.
[http://dx.doi.org/10.1016/j.actbio.2009.08.008] [PMID: 19671456]

[36]   D. Chimene, K.K. Lennox, R.R. Kaunas, and A.K. Gaharwar, "Advanced bioinks for 3D printing: A materials science perspective", *Ann. Biomed. Eng.,* vol. 44, no. 6, pp. 2090-2102, 2016.
[http://dx.doi.org/10.1007/s10439-016-1638-y] [PMID: 27184494]

[37]   J. Yu, S.A. Park, W.D. Kim, T. Ha, Y.Z. Xin, J. Lee, and D. Lee, "Current advances in 3D bioprinting technology and its applications for tissue engineering polymers", *Polymers (Basel),* vol. 12, no. 12, p. 2958, 2020.
[http://dx.doi.org/10.3390/polym12122958] [PMID: 33322291]

[38]   M. Guvendiren, J. Molde, R.M.D. Soares, and J. Kohn, "Designing biomaterials for 3D printing", *ACS Biomater. Sci. Eng.,* vol. 2, no. 10, pp. 1679-1693, 2016.
[http://dx.doi.org/10.1021/acsbiomaterials.6b00121] [PMID: 28025653]

[39]   S. Ji, and M. Guvendiren, "Recent advances in bioink design for 3D bioprinting of tissues and organs", *Front. Bioeng. Biotechnol.,* vol. 5, p. 23, 2017.
[http://dx.doi.org/10.3389/fbioe.2017.00023] [PMID: 28424770]

[40]   V. Carvalho, I. Gonçalves, T. Lage, R.O. Rodrigues, G. Minas, S.F.C.F. Teixeira, A.S. Moita, T. Hori, H. Kaji, and R.A. Lima, "3D printing techniques and their applications to organ-on-a-chip platforms: A systematic review", *Sensors (Basel),* vol. 21, no. 9, p. 3304, 2021.
[http://dx.doi.org/10.3390/s21093304] [PMID: 34068811]

[41]   R. Klebe, "Cytoscribing: A method for micropositioning cells and the construction of two- and three-dimensional synthetic tissues", *Exp. Cell Res.,* vol. 179, no. 2, pp. 362-373, 1988.
[http://dx.doi.org/10.1016/0014-4827(88)90275-3] [PMID: 3191947]

[42]   H. Cui, M. Nowicki, J.P. Fisher, and L.G. Zhang, "3D Bioprinting for Organ Regeneration", *Adv. Healthc. Mater.,* vol. 6, no. 1, p. 1601118, 2017.
[http://dx.doi.org/10.1002/adhm.201601118] [PMID: 27995751]

[43]   T. Xu, W. Zhao, J.M. Zhu, M.Z. Albanna, J.J. Yoo, and A. Atala, "Complex heterogeneous tissue constructs containing multiple cell types prepared by inkjet printing technology", *Biomaterials,* vol. 34, no. 1, pp. 130-139, 2013.
[http://dx.doi.org/10.1016/j.biomaterials.2012.09.035] [PMID: 23063369]

[44]   J. Malda, J. Visser, F.P. Melchels, T. Jüngst, W.E. Hennink, W.J.A. Dhert, J. Groll, and D.W. Hutmacher, "25th anniversary article: Engineering hydrogels for biofabrication", *Adv. Mater.,* vol. 25, no. 36, pp. 5011-5028, 2013.
[http://dx.doi.org/10.1002/adma.201302042] [PMID: 24038336]

[45]   X. Cui, T. Boland, D.D. D'Lima, and M.K. Lotz, "Thermal inkjet printing in tissue engineering and regenerative medicine", *Recent Pat. Drug Deliv. Formul.,* vol. 6, no. 2, pp. 149-155, 2012.
[http://dx.doi.org/10.2174/187221112800672949] [PMID: 22436025]

[46]   X. Cui, and T. Boland, "Human microvasculature fabrication using thermal inkjet printing technology", *Biomaterials,* vol. 30, no. 31, pp. 6221-6227, 2009.
[http://dx.doi.org/10.1016/j.biomaterials.2009.07.056] [PMID: 19695697]

[47]   R. Landers, U. Hübner, R. Schmelzeisen, and R. Mülhaupt, "Rapid prototyping of scaffolds derived from thermoreversible hydrogels and tailored for applications in tissue engineering", *Biomaterials,* vol. 23, no. 23, pp. 4437-4447, 2002.
[http://dx.doi.org/10.1016/S0142-9612(02)00139-4] [PMID: 12322962]

[48]   A.D. Graham, S.N. Olof, M.J. Burke, J.P.K. Armstrong, E.A. Mikhailova, J.G. Nicholson, S.J. Box, F.G. Szele, A.W. Perriman, and H. Bayley, "High-resolution patterned cellular constructs by droplet-based 3D printing", *Sci. Rep.,* vol. 7, no. 1, p. 7004, 2017.
[http://dx.doi.org/10.1038/s41598-017-06358-x] [PMID: 28765636]

[49]   J.M. Unagolla, and A.C. Jayasuriya, "Hydrogel-based 3D bioprinting: A comprehensive review on

cell-laden hydrogels, bioink formulations, and future perspectives", *Appl. Mater. Today,* vol. 18, p. 100479, 2020.
[http://dx.doi.org/10.1016/j.apmt.2019.100479] [PMID: 32775607]

[50] Z. Gu, J. Fu, H. Lin, and Y. He, "Development of 3D bioprinting: From printing methods to biomedical applications", *Asian J. of Pharm. Sci.,* vol. 15, no. 5, pp. 529-557, 2020.
[http://dx.doi.org/10.1016/j.ajps.2019.11.003] [PMID: 33193859]

[51] X. Lan, Y. Liang, E.J.N Erkut, M. Kunze, A. Mulet-Sierra, T. Gong, M. Osswald, K. Ansari, H. Seikaly, Y. Boluk, and A.B. Adesida, "Bioprinting of human nasoseptal chondrocytes-laden collagen hydrogel for cartilage tissue engineering", *The FASEB Journal.,* vol. 35, pp. 1-14, 2021.
[http://dx.doi.org/10.1096/fj.202002081R]

[52] X. Lan, Y. Liang, M. Vyhlidal, E.J.N. Erkut, M. Kunze, A. Mulet-Sierra, M. Osswald, K. Ansari, H. Seikaly, Y. Boluk, and A.B. Adesida, "*In vitro* maturation and *in vivo* stability of bioprinted human nasal cartilage", *J. Tissue Eng.,* vol. 13, 2022.
[http://dx.doi.org/10.1177/20417314221086368] [PMID: 35599742]

[53] K. Glover, E. Mathew, G. Pitzanti, E. Magee, and D.A. Lamprou, "3D bioprinted scaffolds for diabetic wound-healing applications", *Drug Deliv. Transl. Res.,* 2022.
[http://dx.doi.org/10.1007/s13346-022-01115-8] [PMID: 35018558]

[54] I.T. Ozbolat, and M. Hospodiuk, "Current advances and future perspectives in extrusion-based bioprinting", *Biomaterials,* vol. 76, pp. 321-343, 2016.
[http://dx.doi.org/10.1016/j.biomaterials.2015.10.076] [PMID: 26561931]

[55] V. Mironov, T. Trusk, V. Kasyanov, S. Little, R. Swaja, and R. Markwald, "Biofabrication: a 21st century manufacturing paradigm", *Biofabrication,* vol. 1, no. 2, p. 022001, 2009.
[http://dx.doi.org/10.1088/1758-5082/1/2/022001] [PMID: 20811099]

[56] F. You, B.F. Eames, and X. Chen, "Application of extrusion-based hydrogel bioprinting for cartilage tissue engineering", *Int. J. Mol. Sci.,* vol. 18, no. 7, p. 1597, 2017.
[http://dx.doi.org/10.3390/ijms18071597] [PMID: 28737701]

[57] D.J. Odde, and M.J. Renn, "Laser-guided direct writing for applications in biotechnology", *Trends Biotechnol.,* vol. 17, no. 10, pp. 385-389, 1999.
[http://dx.doi.org/10.1016/S0167-7799(99)01355-4] [PMID: 10481169]

[58] F. Guillemot, A. Souquet, S. Catros, and B. Guillotin, "Laser-assisted cell printing: principle, physical parameters versus cell fate and perspectives in tissue engineering", *Nanomedicine (Lond.),* vol. 5, no. 3, pp. 507-515, 2010.
[http://dx.doi.org/10.2217/nnm.10.14] [PMID: 20394540]

[59] E.S. Bishop, S. Mostafa, M. Pakvasa, H.H. Luu, M.J. Lee, J.M. Wolf, G.A. Ameer, T.C. He, and R.R. Reid, "3-D bioprinting technologies in tissue engineering and regenerative medicine: Current and future trends", *Genes Dis.,* vol. 4, no. 4, pp. 185-195, 2017.
[http://dx.doi.org/10.1016/j.gendis.2017.10.002] [PMID: 29911158]

[60] T. Agarwal, I. Chiesa, D. Presutti, V. Irawan, K.Y. Vajanthri, M. Costantini, Y. Nakagawa, S.A. Tan, P. Makvandi, E.N. Zare, E. Sharifi, C. De Maria, T. Ikoma, and T.K. Maiti, "Recent advances in bioprinting technologies for engineering different cartilage-based tissues", *Mater. Sci. Eng. C,* vol. 123, p. 112005, 2021.
[http://dx.doi.org/10.1016/j.msec.2021.112005] [PMID: 33812625]

[61] Y.J. Seol, H.W. Kang, S.J. Lee, A. Atala, and J.J. Yoo, "Bioprinting technology and its applications", *Eur. J. Cardiothorac. Surg.,* vol. 46, no. 3, pp. 342-348, 2014.
[http://dx.doi.org/10.1093/ejcts/ezu148] [PMID: 25061217]

[62] D.V. Leonov, Y.A. Spirina, A.A. Yatsenko, V.A. Kushnarev, E.M. Ustinov, and S.V. Barannikov, "Advanced 3D bioprinting technologies", *Cell Tissue Biol.,* vol. 15, no. 6, pp. 616-627, 2021.
[http://dx.doi.org/10.1134/S1990519X21060134]

[63]   L.S.S.M. Magalhães, F.E.P. Santos, C.M.V. Elias, S. Afewerki, G.F. Sousa, A.S.A. Furtado, F.R. Marciano, and A.O. Lobo, "Printing 3D hydrogel structures employing low-cost stereolithography technology", *J. Funct. Biomater.,* vol. 11, no. 1, p. 12, 2020.
[http://dx.doi.org/10.3390/jfb11010012] [PMID: 32098431]

[64]   V.B. Morris, S. Nimbalkar, M. Younesi, P. McClellan, and O. Akkus, "Mechanical properties, cytocompatibility and manufacturability of chitosan: PEGDA hybrid-gel scaffolds by stereolithography", *Ann. Biomed. Eng.,* vol. 45, no. 1, pp. 286-296, 2017.
[http://dx.doi.org/10.1007/s10439-016-1643-1] [PMID: 27164837]

[65]   U. Demirci, and G. Montesano, "Single cell epitaxy by acoustic picolitre droplets", *Lab Chip,* vol. 7, no. 9, pp. 1139-1145, 2007.
[http://dx.doi.org/10.1039/b704965j] [PMID: 17713612]

[66]   H. Gudapati, M. Dey, and I. Ozbolat, "A comprehensive review on droplet-based bioprinting: Past, present and future", *Biomaterials,* vol. 102, pp. 20-42, 2016.
[http://dx.doi.org/10.1016/j.biomaterials.2016.06.012] [PMID: 27318933]

[67]   T. Xu, K.W. Binder, M.Z. Albanna, D. Dice, W. Zhao, J.J. Yoo, and A. Atala, "Hybrid printing of mechanically and biologically improved constructs for cartilage tissue engineering applications", *Biofabrication,* vol. 5, no. 1, p. 015001, 2012.
[http://dx.doi.org/10.1088/1758-5082/5/1/015001] [PMID: 23172542]

[68]   M.S. Kim, and G. Kim, "Three-dimensional electrospun polycaprolactone (PCL)/alginate hybrid composite scaffolds", *Carbohydr. Polym.,* vol. 114, pp. 213-221, 2014.
[http://dx.doi.org/10.1016/j.carbpol.2014.08.008] [PMID: 25263884]

[69]   S. Catros, F. Guillemot, A. Nandakumar, S. Ziane, L. Moroni, P. Habibovic, C. van Blitterswijk, B. Rousseau, O. Chassande, J. Amédée, and J.C. Fricain, "Layer-by-layer tissue microfabrication supports cell proliferation *in vitro* and *in vivo*", *Tissue Eng. Part C Methods,* vol. 18, no. 1, pp. 62-70, 2012.
[http://dx.doi.org/10.1089/ten.tec.2011.0382] [PMID: 21895563]

[70]   B.S. Kim, J.S. Lee, G. Gao, and D.W. Cho, "Direct 3D cell-printing of human skin with functional transwell system", *Biofabrication,* vol. 9, no. 2, p. 025034, 2017.
[http://dx.doi.org/10.1088/1758-5090/aa71c8] [PMID: 28586316]

[71]   M.R. Khosravani, and T. Reinicke, "3D-printed sensors: Current progress and future challenges", *Sens. Actuators A Phys.,* vol. 305, p. 111916, 2020.
[http://dx.doi.org/10.1016/j.sna.2020.111916]

[72]   C. Faccini de Lima, L.A. van der Elst, V.N. Koraganji, M. Zheng, M. Gokce Kurtoglu, and A. Gumennik, "Towards digital manufacturing of smart multimaterial fibers", *Nanoscale Res. Lett.,* vol. 14, no. 1, p. 209, 2019.
[http://dx.doi.org/10.1186/s11671-019-3031-x] [PMID: 31214792]

[73]   E. Ferrari, C. Palma, S. Vesentini, P. Occhetta, and M. Rasponi, "Integrating biosensors in organs-on-chip devices: A perspective on current strategies to monitor microphysiological systems", *Biosensors (Basel),* vol. 10, no. 9, p. 110, 2020.
[http://dx.doi.org/10.3390/bios10090110] [PMID: 32872228]

[74]   R.O. Rodrigues, P.C. Sousa, J. Gaspar, M. Bañobre-López, R. Lima, and G. Minas, "Organ-on-a-chip: A preclinical microfluidic platform for the progress of nanomedicine", *Small,* vol. 16, no. 51, p. 2003517, 2020.
[http://dx.doi.org/10.1002/smll.202003517] [PMID: 33236819]

[75]   S. Jalili-Firoozinezhad, C.C. Miranda, and J.M.S. Cabral, "Modeling the human body on microfluidic chips", *Trends Biotechnol.,* vol. 39, no. 8, pp. 838-852, 2021.
[http://dx.doi.org/10.1016/j.tibtech.2021.01.004] [PMID: 33581889]

[76]   R. Rebelo, A.I. Barbosa, D. Caballero, I.K. Kwon, J.M. Oliveira, S.C. Kundu, R.L. Reis, and V.M.

Correlo, "3D biosensors in advanced medical diagnostics of high mortality diseases", *Biosens. Bioelectron.,* vol. 130, pp. 20-39, 2019.
[http://dx.doi.org/10.1016/j.bios.2018.12.057] [PMID: 30716590]

[77] M.S. Mannoor, Z. Jiang, T. James, Y.L. Kong, K.A. Malatesta, W.O. Soboyejo, N. Verma, D.H. Gracias, and M.C. McAlpine, "3D printed bionic ears", *Nano Lett.,* vol. 13, no. 6, pp. 2634-2639, 2013.
[http://dx.doi.org/10.1021/nl4007744] [PMID: 23635097]

[78] H. Mao, L. Yang, H. Zhu, L. Wu, P. Ji, J. Yang, and Z. Gu, "Recent advances and challenges in materials for 3D bioprinting", *Prog. Nat. Sci.,* vol. 30, no. 5, pp. 618-634, 2020.
[http://dx.doi.org/10.1016/j.pnsc.2020.09.015]

[79] H. Li, Y.J. Tan, K.F. Leong, and L. Li, "3D bioprinting of highly thixotropic alginate/methylcellulose hydrogel with strong interface bonding", *ACS Appl. Mater. Interfaces,* vol. 9, no. 23, pp. 20086-20097, 2017.
[http://dx.doi.org/10.1021/acsami.7b04216] [PMID: 28530091]

[80] B. Duan, E. Kapetanovic, L.A. Hockaday, and J.T. Butcher, "Three-dimensional printed trileaflet valve conduits using biological hydrogels and human valve interstitial cells", *Acta Biomater.,* vol. 10, no. 5, pp. 1836-1846, 2014.
[http://dx.doi.org/10.1016/j.actbio.2013.12.005] [PMID: 24334142]

[81] A. Kjar, B. McFarland, K. Mecham, N. Harward, and Y. Huang, "Engineering of tissue constructs using coaxial bioprinting", *Bioact. Mater.,* vol. 6, no. 2, pp. 460-471, 2021.
[http://dx.doi.org/10.1016/j.bioactmat.2020.08.020] [PMID: 32995673]

[82] G. Gao, J.H. Lee, J. Jang, D.H. Lee, J.S. Kong, B.S. Kim, Y.J. Choi, W.B. Jang, Y.J. Hong, S.M. Kwon, and D.W. Cho, "Tissue engineered bio-blood-vessels constructed using a tissue-specific bioink and 3D coaxial cell printing technique: A novel therapy for ischemic disease", *Adv. Funct. Mater.,* vol. 27, no. 33, p. 1700798, 2017.
[http://dx.doi.org/10.1002/adfm.201700798]

[83] M.H. Kim, and S.Y. Nam, "Assessment of coaxial printability for extrusion-based bioprinting of alginate-based tubular constructs", *Bioprinting,* vol. 20, p. e00092, 2020.
[http://dx.doi.org/10.1016/j.bprint.2020.e00092]

[84] S. Hong, J.S. Kim, B. Jung, C. Won, and C. Hwang, "Coaxial bioprinting of cell-laden vascular constructs using a gelatin–tyramine bioink", *Biomater. Sci.,* vol. 7, no. 11, pp. 4578-4587, 2019.
[http://dx.doi.org/10.1039/C8BM00618K] [PMID: 31433402]

[85] L. Shao, Q. Gao, C. Xie, J. Fu, M. Xiang, and Y. He, "Bioprinting of Cell–Laden Microfiber: Can It Become a Standard Product?", *Adv. Healthc. Mater.,* vol. 8, no. 9, p. 1900014, 2019.
[http://dx.doi.org/10.1002/adhm.201900014] [PMID: 30866173]

[86] T.J. Hinton, Q. Jallerat, R.N. Palchesko, J.H. Park, M.S. Grodzicki, H.J. Shue, M.H. Ramadan, A.R. Hudson, and A.W. Feinberg, "Three-dimensional printing of complex biological structures by freeform reversible embedding of suspended hydrogels", *Sci. Adv.,* vol. 1, no. 9, p. e1500758, 2015.
[http://dx.doi.org/10.1126/sciadv.1500758] [PMID: 26601312]

[87] Q. Chen, H. Chen, L. Zhu, and J. Zheng, "Fundamentals of double network hydrogels", *J. Mater. Chem. B Mater. Biol. Med.,* vol. 3, no. 18, pp. 3654-3676, 2015.
[http://dx.doi.org/10.1039/C5TB00123D] [PMID: 32262840]

[88] T.C. Suekama, J. Hu, T. Kurokawa, J.P. Gong, and S.H. Gehrke, "Double-network strategy improves fracture properties of chondroitin sulphate networks", *ACS Macro Lett.,* vol. 2, no. 2, pp. 137-140, 2013.
[http://dx.doi.org/10.1021/mz3006318] [PMID: 35581775]

[89] Q. Chen, L. Zhu, L. Huang, H. Chen, K. Xu, Y. Tan, P. Wang, and J. Zheng, "Fracture of the physically cross-linked first network in hybrid double network hydrogels", *Macromolecules,* vol. 47, no. 6, pp. 2140-2148, 2014.

[http://dx.doi.org/10.1021/ma402542r]

[90]   S.E. Bakarich, R. Gorkin III, M. Panhuis, and G.M. Spinks, "4D printing with mechanically robust, thermally actuating hydrogels", *Macromol. Rapid Commun.,* vol. 36, no. 12, pp. 1211-1217, 2015.
[http://dx.doi.org/10.1002/marc.201500079] [PMID: 25864515]

[91]   M.K. Jaiswal, J.R. Xavier, J.K. Carrow, P. Desai, D. Alge, and A.K. Gaharwar, "Mechanically stiff nanocomposite hydrogels at ultralow nanoparticle content", *ACS Nano,* vol. 10, no. 1, pp. 246-256, 2016.
[http://dx.doi.org/10.1021/acsnano.5b03918] [PMID: 26670176]

[92]   P. Kerativitayanan, J.K. Carrow, and A.K. Gaharwar, "Nanomaterials for engineering stem cell responses", *Adv. Healthc. Mater.,* vol. 4, no. 11, pp. 1600-1627, 2015.
[http://dx.doi.org/10.1002/adhm.201500272] [PMID: 26010739]

[93]   T. Thakur, J.R. Xavier, L. Cross, M.K. Jaiswal, E. Mondragon, R. Kaunas, and A.K. Gaharwar, "Photocrosslinkable and elastomeric hydrogels for bone regeneration", *J. Biomed. Mater. Res. A,* vol. 104, no. 4, pp. 879-888, 2016.
[http://dx.doi.org/10.1002/jbm.a.35621] [PMID: 26650507]

[94]   A.K. Gaharwar, N.A. Peppas, and A. Khademhosseini, "Nanocomposite hydrogels for biomedical applications", *Biotechnol. Bioeng.,* vol. 111, no. 3, pp. 441-453, 2014.
[http://dx.doi.org/10.1002/bit.25160] [PMID: 24264728]

[95]   K. Markstedt, A. Mantas, I. Tournier, H. Martínez Ávila, D. Hägg, and P. Gatenholm, "H.C. Martı'nez A´ vila, D. Ha¨gg, and P. Gatenholm, "3D Bioprinting human chondrocytes with nanocellulose-alginate bioink for cartilage tissue engineering applications,"", *Biomacromolecules,* vol. 16, no. 5, pp. 1489-1496, 2015.
[http://dx.doi.org/10.1021/acs.biomac.5b00188] [PMID: 25806996]

[96]   C.B. Highley, C.B. Rodell, and J.A. Burdick, "Direct 3D printing of shear-thinning hydrogels into self-healing hydrogels", *Adv. Mater.,* vol. 27, no. 34, pp. 5075-5079, 2015.
[http://dx.doi.org/10.1002/adma.201501234] [PMID: 26177925]

[97]   M. Kesti, M. Müller, J. Becher, M. Schnabelrauch, M. D'Este, D. Eglin, and M. Zenobi-Wong, "A versatile bioink for three-dimensional printing of cellular scaffolds based on thermally and photo-triggered tandem gelation", *Acta Biomater.,* vol. 11, pp. 162-172, 2015.
[http://dx.doi.org/10.1016/j.actbio.2014.09.033] [PMID: 25260606]

[98]   F.L.C. Morgan, L. Moroni, and M.B. Baker, "Dynamic Bioinks to Advance Bioprinting", *Adv. Healthc. Mater.,* vol. 9, no. 15, p. 1901798, 2020.
[http://dx.doi.org/10.1002/adhm.201901798] [PMID: 32100963]

[99]   B.M. Richardson, D.G. Wilcox, M.A. Randolph, and K.S. Anseth, "Hydrazone covalent adaptable networks modulate extracellular matrix deposition for cartilage tissue engineering", *Acta Biomater.,* vol. 83, pp. 71-82, 2019.
[http://dx.doi.org/10.1016/j.actbio.2018.11.014] [PMID: 30419278]

[100] L.L. Wang, C.B. Highley, Y-C. Yeh, J.H. Galarraga, S. Uman, and J.A. Burdick, ""3D extrusion bioprinting of single- and double-network hydrogels containing dynamic covalent crosslinks" J", *Biomed. Mater. Res. Part A,* vol. 106, no. 4, pp. 865-875, 2018.
[http://dx.doi.org/10.1002/jbm.a.36323] [PMID: 29314616]

[101] T. Hu, X. Cui, M. Zhu, M. Wu, Y. Tian, B. Yao, W. Song, Z. Niu, S. Huang, and X. Fu, "3D-printable supramolecular hydrogels with shear-thinning property: fabricating strength tunable bioink *via* dual crosslinking", *Bioact. Mater.*, vol. 5, no. 4, pp. 808-818, 2020.
[http://dx.doi.org/10.1016/j.bioactmat.2020.06.001] [PMID: 32637745]

[102] A. Nademezhad, O.S. Caliskan, F. Topuz, F. Afghah, B. Erman, and B. Koc, "Nanocomposite bioinks based on agarose and 2D nanosilicates with tunable flow properties and bioactivity for 3D bioprinting", *ACS Appl. Bio Mater.*, vol. 2, no. 2, pp. 796-806, 2019.
[http://dx.doi.org/10.1021/acsabm.8b00665] [PMID: 35016284]

Biosensors for Neurodegenerative Diseases

Gaurav Mishra[1,*], Anand Maurya[1], Anurag Kumar Singh[2], Marjan Talebi[3], Rajendra Awasthi[4] and Manmath Kumar Nandi[1]

[1] *Institute of Medical Sciences, Faculty of Ayurveda, Department of Medicinal Chemistry, Banaras Hindu University, Varanasi, 221005, Uttar Pradesh, India*

[2] *Centre of Experimental Medicine & Surgery, Institute of Medical Sciences, Banaras Hindu University, Varanasi 221005, Uttar Pradesh, India*

[3] *Department of Pharmacognosy, School of Pharmacy, Shahid Beheshti University of Medical Sciences, Tehran, Iran*

[4] *Department of Pharmaceutical Sciences, School of Health Sciences and Technology, University of Petroleum and Energy Studies (UPES), Energy Acres, Bidholi, Via-Prem Nagar, Dehradun – 248 007, Uttarakhand, India*

Abstract: Since the conception of biosensor technology in biomedical research, this field is emerging as a promising and high-throughput tool for neuro-engineering and neurosciences research. It has been postulated that the accumulating property proteins are the basic cause of neurodegenerative diseases, such as Parkinson's disease, Alzheimer's disease and prion diseases. Thus, neurodegenerative diseases are also called "protein misfolding disorders". Biosensors have a wide range of applications in biomedical research, including optical and electrochemical detection of biometal-protein interactions, detection of biomarkers, such as β-amyloids, apolipoprotein, and tau proteins, and microRNA in blood and cerebrospinal fluid in neurodegenerative diseases. These are composed of primary biological recognition elements that convert the chemical signal into the voltage or current that evaluates the physical signal by preparing a plot of sensor response against the analyte concentration. This chapter presents a bird's eye view on various aspects of progress in biosensor development with special emphasis on their application, including metal-protein interactions studies, detection of neurotransmitters using aptamers and calixarenes, detection of biomarkers proteins, such as α-synuclein for Parkinson's disease, apolipoprotein, tau and β-amyloid proteins for Alzheimer's disease, and prion proteins. The chapter also summarizes the novel materials reported for improved biosensor performance. This chapter will be of high relevance to the biological scientists working in neuro-engineering and neurosciences research.

* **Corresponding Author Gaurav Mishra**: Institute of Medical Sciences, Faculty of Ayurveda, Department of Medicinal Chemistry, Banaras Hindu University, Varanasi, 221005, Uttar Pradesh, India; Email: vatsgaurav880@gmail.com

Keywords: Alzheimer's disease, Biomarker, Biosensors, Neuro-biosensors, Neurodegenerative diseases, Nanomaterials, Neurocognitive disorders, Nanotechnology, Parkinson's disease.

BACKGROUND

Early diagnosis of neurodegeneration-focused illnesses that worsen with advancing age is essential because they have an impact on the patient's quality of life and medication. Two well-known examples of neurodegeneration in Parkinson's and Alzheimer's are characterized by dementia and nerve cell death. The identification of biomarkers-recognizable substances in bodily fluids involved in the onset or development of neurodegenerative manifestations could make an early diagnosis of neurodegenerative, demyelinating illnesses and autism spectrum disorder possible.

In order to more accurately identify potential biomarkers of the neurodegenerative process and explore the disease's state prognosis, research on biosensors has increased with the objective to identify the target analyte having maximum specificity. This book chapter aims to give a summary of the neuro-biosensors developed based on biomarkers found in body fluids for the identification and management of illnesses regarding the degeneration of neurons with preference to most common Parkinson's disease (PD) and Alzheimer's disease (AD), as well as a summary of the neurodegeneration pathway, which may be directly linked to early diagnosis of the disease. It is an urgent requirement for the treatment. Practically, neuro-biosensing technologies make a connection between various scientific disciplines, such as neuroscience, pharmacology, nanotechnology, electrochemistry, photometry and electronics.

Over the past decades, biosensors identified by using optical and electrochemical technologies, are at the forefront because of developments in material science, such as aptasensors, exosomes, nanomaterials, and wearable biosensors. In the current study, the most recent advances in biosensors used for the detection of neurodegeneration based on optical and electrochemical techniques are evaluated in all aspects.

INTRODUCTION

Neurodegenerative disorders (NDDs), also known as "protein misfolding disorders," are among the most life-limiting diseases associated with significant deficits and impairments among the many age-related ailments that people experience [1].

Neuronal death is caused by various reasons, including excitotoxicity, mitochondrial malfunction, and extracellular buildup of dangerous chemicals. This may affect both physical and mental functioning [2]. Thus, the prognosis of the disease's progress or the effectiveness of the treatment may be crucial for enhancing the quality of life of the affected person.

Sophisticated neuro-biosensing technologies have been developed as a result of the identification of biomarkers in biofluids for the detection of neurodegenerative illnesses, such as Alzheimer's disease (AD) (a later life disease occurrence, categorized as the most common type of dementia), Parkinson's disease (PD) (a demyelinating disorder that affects a body part having the inability to move, as seen in multiple sclerosis (MS) [3, 4], prion diseases, and Huntington's disease (HD) [5]. The need for early diagnosis is also emphasized in the whole pathophysiology of the neurodegenerative pathway [6].

Applications for neuro-biosensors mainly center on the detection of amyloid-beta peptides [7], tau proteins, reactive oxygen species (ROS), lactoferrin, acetylcholine [8], miRNAs [9], transition metals [10], epigenetics, and apolipoprotein in AD [11, 12], dopamine, urate, ascorbic acid, 8-hydroxy-2' - deoxyguanosine, apolipoproteins (A1 and E), DJ-1, and alpha-synuclein in PD [13], immune cells and inflammatory markers in Amyotrophic lateral sclerosis (ALS) [14], a protein in human HD (huntingtin) and in prion diseases (prion) and also demyelinating disorders, such as MS [15, 16].

Despite having a lengthy history, bioanalytical approaches have just achieved an important milestone. Biosensors are examples of analytical machinery that transform biological reactions into quantitative signals. To locate or assist in diagnosing NDDs biomarkers and also autism spectrum disorder (ASD) [17], biosensors offer a sensitive, targeted, specific, and reasonably priced option [14, 18]. Biosensors are used to advance drug development, observe an illness, and detect several substances, including disease-causing biomarkers [19].

APPLICATIONS OF BIOSENSORS IN NEURODEGENERATIVE DISEASES

The recognition layer of a biosensor is made up of biological components that connect with the targeted analyzers, such as nucleic acids (DNA & RNA), peptides, antibodies, antigens and enzymes [20]. Additionally, unique bio-sample based technologies will be used in biosensors with sizes ranging from micro to femto, including biosensors based on various biofluids such as Cerebrospinal fluid (CSF), urine, blood, tear, and saliva [21]. Biosensors can be classified into different groups on the basis of signal types they receive, including optical, electrochemical, plasmonic, thermometric, field-effect transistor-based sensor configurations, and piezoelectric biosensors. Modern technologies employ nanomaterial-based biosensors to address problems with conventional forms, including selectivity and sensitivity [22].

Some research teams have concentrated on developing biosensors in CSF fluids for the recognition of neurodegeneration, which establishes the gold standard of traditional clinical assessments, despite the invasiveness of the sampling technique. The available literature for research on biosensors based on human serum or plasma has been validated clinically for NDDs [23]. The developed biosensors have occasionally been tested first in an environmental condition that is artificially designed before being used with a real biofluid in a clinical context [24]. Aptamers are efficient biorecognition tools in the case of NDDs diagnosis [25]. Metal-based nano-sensors, carbon nanostructures, and further nanotech-nology-based biosensing equipment are used in the diagnosis of NDDs [26]. Alpha-synuclein, a PD biomarker, was detected by the electrochemical sensor based on gold nanorods (Au-NRs) is a recently example. Since serum does not contain enough of the biomarker to accurately identify the condition, they came to the conclusion that alternate fluids, such as saliva, should be examined for alpha-synuclein [27]. The brain-barrier performance with respect to new therapies, which is controlled by blood - the brain barrier, might be a very helpful tool for drug permeability testing in addition to offering a deeper knowledge of NDDs. A highly attractive development of BBB-OoC technology is the inclusion of detecting systems to offer real-time monitoring of biomarkers continuously and a completely automated study of drug permeability, rendering more effective platforms for the development of new therapies for commercial purposes [28].

WEARABLE BIOSENSORS IN NDDS

Recent years have seen a rapid uptake of wearable sensor technologies by a variety of conventional consumers and manufacturers of medical products. Wearable sensors are rapidly expanding due to a number of factors, including advancements in small electronic devices at both economics and ergonomics levels, the prevalence of smartphones and connected devices, the rising of health awareness of consumers, and the requirement for physicians to continuously collect medical data of high-quality from their patients in real-time observative data [29]. All across the body, notably in the regions below the feet, waist, hands and arms, are exocrine glands, which secrete body fluids. The newest wearable electrochemical biosensors are personalized for the screening of body fluids comprising sweat, tears and interstitial fluid [30]. As a result, it is gaining popularity as a source of information on pharmaceuticals and hydration levels. Recently, body fluid sweat has been used to detect sickness indications and expose the biomarkers associated with hemodynamic and metabolic states like salt, caffeine, potassium, lactate, calcium, phosphate, oxygen, ethanol, and glucose [31]. Integration of electrochemical sensing systems with modern, covert body-worn inertial sensors is a significant potential possibility for the use of biosensors. Wearable inertial sensors, such as triaxial accelerometers and gyroscopes, can be mounted to various body areas depending on the motor task being researched. Inertial sensors have the ability to record motion and spatiotemporal data, which may be utilized to collect critical data on motor function [32].

The use of wearable biosensor technology might be a useful technique to assess the biochemical-physiological changes and digital biomarkers linked to NDDs [33]. It may be able to track pre-clinical AD-related cognitive changes using longitudinal sleep and heart rate variability data collection, entirely individualized therapy [34], and machine-human interaction [35, 36]. Besides, in some studies, the role of wearable electrochemical biosensors has been stated regarding the diagnosis and follow-up, including multidimensional circadian monitoring in PD patients. For instance, a microneedle-constructed interstitial fluid sensor implantable biosensor [37] is principally tailor-fit to determine the quantity of levodopa in an unremitting and direct matter in interstitial fluid, which is useful for patients with PD [30].

BIOSENSORS FOR DETECTION

Many biological reactions occur inside our body, and biosensors change the chemical signal into an electrical signal that can be easily measurable and reproducible. These devices have a lot of promise in the fields of the environment [38 - 40], safety [41, 42] and sports [43, 44] respectively. Due to low cost, simple and rapid analysis, these devices are promising tools for the diagnosis and management of various diseases [45 - 48]. Since the 1960s, these biosensors are contributing promising results after innovation done by Clark and Lyons [49]. Further, modification and development in this field lead to improved materials, new constructions and detection approaches, and more stable biorecognition elements, making the data generation more reproducible and reliable.

Alzheimer's Diseases (AD) detection

There are different ways to diagnose AD, such as the Cerebral fluid analysis method, Cognitive analysis. The more renowned and frequent method used for medical imaging is magnetic resonance imaging (MRI). The evaluation of atrophy within the hippocampus is generally found in the brain *via* positron emission tomography (PET). For the assessment of Aβ accumulation within the cortex. Thus results need biosensors, *i.e.*, Point-of-Care. Therefore, these devices are intended for *in situ* performance within limited resources and options. Today, several Point-of-Care biosensors are useful to patients and doctors and make it convenient for both.

The biosensor application brings many enormous changes in AD detection. One of the major benefits is the low cost of physiological analysis and neuroimaging methods. The second benefit is the non-invasive technique so that it is highly patient compliance and they can find biomarkers in different biofluids compared to CSF. Therefore a lumber puncture is necessary to collect CSF samples; patient compliance is essential. This is a very painful method, so it is rarely used and more repletion of the method brings biomarker levels affected; disease monitoring will also be affected. As we see, AD is more commonly found in older patients, so these people make CSF less suitable for the analysis. The neuroimaging technique of independent single people is very costly, and it is also limited to the diagnosis of very less patients.

Biosensors are very useful in diagnosis in the initial stage of the patient, at the time when there is no change in the structure of the brain. Since their use relies on biomarkers detection, which is only possible in the initial stage of the disease. So this study is more useful and impactful to more patients for early diagnosis and further treatment. Therefore also provide a platform to facilitate new findings and better management of diseases Table (**1**).

In developing the biosensor, an optical biosensor emerges in AD diagnosis and treatment, *i.e.*, Palladino *et al.* [50] proposed SPR to study the formation of Aβ plaque in real-time by immobilizing specific antibodies on the chip of biosensor. The benefit of such a technique is giving way to perform the analysis without fluorophores, which alter the aggregation. This method suggests the investigation regarding the accumulation of inhibitors that help in the development of the treatment of diseases.

Moreira and co-workers [51] proposed an enzymatic biosensor utilized in AchE detection into plasma by using a gold electrode that is highly porous in nature and functionalized with acetylcholinestrases (AChE). The interaction was studied by Square wave voltammetry and cyclic voltammetry, which shows the detection limit of 10 μmol L^{-1} amplification of the signals was done to lower the LOD with the voltammetric method.

An additional emerging technique for developing various biosensing technology for AD is Electrochemical impedance spectroscopy (EIS). With EIS, changes are used to signify biomarker molecules. Therefore, the method's broad acceptance in the literature is a result of its extremely low cost and high sensitivity and low LOD. Rushworth and co-workers [52] prepared a very good and effective EIS-based detection system for Aβ oligomers that employ a fragment of the cellular prion protein as a biorecognition element. Further, a very linear and sharp response up to the detection limit of 0.5 pmol L^{-1} is shown.

Table 1. Techniques applied for AD diagnosis and its characteristics.

–	Neuroimaging Techniques	Biosensors	CSF Analyses (ELISA)
Advantages	1. AD diagnosis with accuracy [50] 2. Stage of sickness and actual condition of the patient determined	1. Patient compliance because applicable to samples such as urine and blood. 2. Different analyses can be done with a single chip, *i.e.*, having a high potential to integrate different steps of analysis.	1. Higher accuracy for AD diagnosis. 2. CSF present diverse, relevant biomarker in AD diagnosis in higher concentration than other biological fluid.
Disadvantages	It is a costly and sophisticated instrument that is less readily available to more population [48].	1. Usually presents complex structures to reach the desired LODs, being hardly constructed in large scales	1. Aspiration of CSF is invasive and painful, being an aggressive procedure for the elderly

Parkinson's Disease (PD) Detection

According to reports, optical biosensors can identify PD markers in human body fluids like blood and CSF very quickly and sensitively. A new biosensor based on QD was developed by Ma *et al.* [53], which was intended for the detection of the mitochondrial complex-I- I abnormalities that were associated with PD.

Ubiquinone-terminated disulphide ligands (QnNS) prepared using a 'Click ' were self-assembled onto the surface of 550 nm emitting Core-shell Cd Se/ZnS QDs. The QD bio-conjugate (QnNS-QDs), when in close vicinity to a properly functioning mitochondrial complex-I, produced fluorescence enhancement, and therefore fluorescence is decreased when any damage occurs to the complex-I. The authors illustrated the fascinating potential of this biosensor for intracellular PD detection using human neural cells (SH-SY5Y). QDs have also been efficaciously used to detect PD biomarkers such as DA. Ankireddy *et al.* [54] reported an indium phosphide/ zinc sulphide (InP/ZnS) based QD to detect DA levels in the attendance of ascorbic acid (AA). A coupling reaction was used to modify the surface of the QDs with L-cysteine and DA was detected by fluorescence quenching of cysteine-capped In P/ZnS QDs, in the presence of AA.

Furthermore, emerging studies recommend that the combined determination of α-synuclein and other biomarkers in CSF display PD-specific patterns, supporting the case for developing multiplex assay strategies that target multiple PD biomarkers [55].

Biosensors in the Current Market and their Utilization

The glucometer is one best example of a biosensor, by which the level of glucose in an individual can be monitored using a drop of blood at their home. Enzymatic biochemical reactions occur inside the glucometer. The electrode of the glucometer immobilizes an enzyme, glucose oxidase, that reacts with blood glucose by using oxygen from the surrounding. The linear correlation of this reaction is read out by the device as an amperometric measurement, which indicates the level of glucose in an individual. A pregnancy test kit and an HIV test kit are two additional examples of biosensor devices used to detect various biomarkers in a person.

CONCLUSION

It is envisaged that the development of neuro-biosensors as advanced diagnostic tools such as MRI and together with the identification and collection of more biomarkers, leads to early diagnosis of neurodegenerative anomalies in an individual, which helps in the treatment and management of diseases with the help of new therapeutic approaches like NDDs. Therefore, advancements in new chips, biosensors are being used to screen and estimate AD indicators such as tau protein.

Further, nanomaterials and biosensor devices are very delicate, more accurate, reliable, and cost-effective. Recently, various biosensors have been available commercially to estimate tau protein with distinctive operating systems that are helpful in effective and new therapeutic approch.

FUTURE REMARKS

It is seen that in last fifty years, the biosensing technique has been rapidly evolving. It offers newer technology that is cost-effective, accurate and sensitive in specific site detection for neurodegenerative biomarkers with its developed substructure. Currently, in view of the glitzy point reached by biosensor technology, these studies are insufficient for now. It is clear that biosensors that will be developed for the detection of multiplex rather than single biomarkers that play a central role in neurodegeneration will have a stronger potential for early diagnosis of the disease.

ABBREVIATIONS

PD	Parkinson's disease
AD	Alzheimer's disease
NDDs	Neurodegenerative disorders
MS	Multiple sclerosis
HD	Huntington's disease
ROS	Reactive oxygen species
ALS	Amyotrophic lateral sclerosis
ASD	Autism spectrum disorder
Au-NRs	Gold nanorods
MRI	Magnetic resonance imaging
PET	Positron emission tomography
CSF	Cerebrospinal fluid
AChE	Acetylcholinestrases
EIS	Electrochemical impedance spectroscopy
QnNS	Ubiquinone-terminated disulphide ligands

ACKNOWLEDGEMENT

Declared none.

REFERENCES

[1]   H.V.S. Ganesh, A.M. Chow, and K. Kerman, "Recent advances in biosensors for neurodegenerative disease detection", *Trends Analyt. Chem.,* vol. 79, no. 79, pp. 363-370, 2016.
[http://dx.doi.org/10.1016/j.trac.2016.02.012]

[2]   M. Talebi, S. İlgün, V. Ebrahimi, M. Talebi, T. Farkhondeh, H. Ebrahimi, and S. Samarghandian, "Zingiber officinale ameliorates Alzheimer's disease and Cognitive Impairments: Lessons from preclinical studies", *Biomed. Pharmacother.,* vol. 133, no. 133, p. 111088, 2021.
[http://dx.doi.org/10.1016/j.biopha.2020.111088] [PMID: 33378982]

[3]   C.M. Abreu, R. Soares-dos-Reis, P.N. Melo, J.B. Relvas, J. Guimarães, M.J. Sá, A.P. Cruz, and I. Mendes Pinto, "Emerging biosensing technologies for neuroinflammatory and neurodegenerative disease diagnostics", *Front. Mol. Neurosci.,* vol. 11, no. 11, p. 164, 2018.
[http://dx.doi.org/10.3389/fnmol.2018.00164] [PMID: 29867354]

[4]   S. Shariati, A. Ghaffarinejad, and E. Omidinia, "Early detection of multiple sclerosis (MS) as a neurodegenerative disease using electrochemical nano-aptasensor", *Microchem. J.,* vol. 178, no. 178, p. 107358, 2022.
[http://dx.doi.org/10.1016/j.microc.2022.107358]

[5]   H. Zetterberg, and B. B. Bendlin, "Neurodegenerative dementias: screening for major threats to healthy longevity with blood biomarkers", *The Lancet Healthy Longevity.,* vol. 2, no. 2, pp. e58-59, 2021.
[http://dx.doi.org/10.1016/S2666-7568(21)00009-X]

[6]   M. Talebi, H. Esmaeeli, M. Talebi, T. Farkhondeh, and S. Samarghandian, "A concise overview of biosensing technologies for the detection of Alzheimer's disease biomarkers", *Curr Pharm Biotechnol.,* vol. 23, no. 5, pp. 634-644, 2022.
[http://dx.doi.org/10.2174/2666796702666210709122407] [PMID: 34250871]

[7]   A. Kaushik, R.D. Jayant, S. Tiwari, A. Vashist, and M. Nair, "Nano-biosensors to detect beta-amyloid for Alzheimer's disease management", *Biosens. Bioelectron.,* vol. 80, no. 80, pp. 273-287, 2016.
[http://dx.doi.org/10.1016/j.bios.2016.01.065] [PMID: 26851586]

[8]   L.C. Brazaca, I. Sampaio, V. Zucolotto, and B.C. Janegitz, "Applications of biosensors in Alzheimer's disease diagnosis", *Talanta,* vol. 210, no. 210, p. 120644, 2020.
[http://dx.doi.org/10.1016/j.talanta.2019.120644] [PMID: 31987214]

[9]   B. Khalilzadeh, M. Rashidi, A. Soleimanian, H. Tajalli, G.S. Kanberoglu, B. Baradaran, and M.R. Rashidi, "Development of a reliable microRNA based electrochemical genosensor for monitoring of miR-146a, as key regulatory agent of neurodegenerative disease", *Int. J. Biol. Macromol.,* vol. 134, no. 134, pp. 695-703, 2019.
[http://dx.doi.org/10.1016/j.ijbiomac.2019.05.061] [PMID: 31082423]

[10]   B. Khalilzadeh, M. Rashidi, A. Soleimanian, H. Tajalli, G.S. Kanberoglu, B. Baradaran, and M.R. Rashidi, "Development of a reliable microRNA based electrochemical genosensor for monitoring of miR-146a, as key regulatory agent of neurodegenerative disease", *Int. J. Biol. Macromol.,* vol. 134, no. 134, pp. 695-703, 2019.
[http://dx.doi.org/10.1016/j.ijbiomac.2019.05.061] [PMID: 31082423]

[11]   S. Li, and K. Kerman, "Electrochemical biosensors for biometal-protein interactions in neurodegenerative diseases", *Biosens. Bioelectron.,* vol. 179, no. 179, p. 113035, 2021.
[http://dx.doi.org/10.1016/j.bios.2021.113035] [PMID: 33578115]

[12]   B. Shui, D. Tao, A. Florea, J. Cheng, Q. Zhao, Y. Gu, W. Li, N. Jaffrezic-Renault, Y. Mei, and Z. Guo, "Biosensors for Alzheimer's disease biomarker detection: A review", *Biochimie,* vol. 147, pp. 13-24, 2018.
[http://dx.doi.org/10.1016/j.biochi.2017.12.015] [PMID: 29307704]

[13]   R. Goldoni, C. Dolci, E. Boccalari, F. Inchingolo, A. Paghi, L. Strambini, D. Galimberti, and G.M. Tartaglia, "Salivary biomarkers of neurodegenerative and demyelinating diseases and biosensors for their detection", *Ageing Res. Rev.,* vol. 76, p. 101587, 2022.
[http://dx.doi.org/10.1016/j.arr.2022.101587] [PMID: 35151849]

[14]   P. Fazlali, A. Mahdian, M. S. Soheilifar, S. M. Amininasab, P. Shafiee, and I. A. Wani, "Nanobiosensors for early detection of neurodegenerative disease", *Journal of Composites and Compounds,* vol. 4, no. 10, pp. 24-36, 2022.
[http://dx.doi.org/10.52547/jcc.4.1.4]

[15]   C.M. Rice, K. Kemp, A. Wilkins, and N.J. Scolding, "Cell therapy for multiple sclerosis: an evolving concept with implications for other neurodegenerative diseases", *Lancet,* vol. 382, no. 9899, pp. 1204-1213, 2013.
[http://dx.doi.org/10.1016/S0140-6736(13)61810-3] [PMID: 24095194]

[16]   H. Xi, and Y. Zhang, "Aptamer detection of neurodegenerative disease biomarkers", In: *Neurodegenerative Diseases Biomarkers.* Humana: New York, NYpp. 361-386. 2022.

[17]   L. Farzin, M. Shamsipur, L. Samandari, and S. Sheibani, "Advances in the design of nanomaterial-based electrochemical affinity and enzymatic biosensors for metabolic biomarkers: A review", *Mikrochim. Acta,* vol. 185, no. 5, p. 276, 2018.
[http://dx.doi.org/10.1007/s00604-018-2820-8] [PMID: 29721621]

[18]   M.N.S. Karaboğa, and M.K. Sezgintürk, "Biosensor approaches on the diagnosis of neurodegenerative diseases: Sensing the past to the future", *J. Pharm. Biomed. Anal.,* vol. 209, no. 209, p. 114479, 2022.
[http://dx.doi.org/10.1016/j.jpba.2021.114479] [PMID: 34861607]

[19]   F.J. Gruhl, B.E. Rapp, and K. Länge, "Biosensors for diagnostic applications. In: Seitz H, Schumacher S, editors", In: *Molecular Diagnostics.* Springer Berlin Heidelberg: Berlin, Heidelberg. 115-48, 2013.

[20]   A. Rezabakhsh, R. Rahbarghazi, and F. Fathi, "Surface plasmon resonance biosensors for detection of Alzheimer's biomarkers; an effective step in early and accurate diagnosis", *Biosens. Bioelectron.,* vol. 167, no. 167, p. 112511, 2020.
[http://dx.doi.org/10.1016/j.bios.2020.112511] [PMID: 32858422]

[21]   C. Toyos-Rodríguez, F.J. García-Alonso, and A. de la Escosura-Muñiz, "Electrochemical biosensors based on nanomaterials for early detection of alzheimer's disease", *Sensors (Basel),* vol. 20, no. 17, p. 4748, 2020.
[http://dx.doi.org/10.3390/s20174748] [PMID: 32842632]

[22]   B.T. Murti, A.D. Putri, Y.J. Huang, S.M. Wei, C.W. Peng, and P.K. Yang, "Clinically oriented Alzheimer's biosensors: expanding the horizons towards point-of-care diagnostics and beyond", *RSC Advances,* vol. 11, no. 33, pp. 20403-20422, 2021.
[http://dx.doi.org/10.1039/D1RA01553B] [PMID: 35479927]

[23]   A. Jeromin, and R. Bowser, "Biomarkers in neurodegenerative diseases", *Adv. Neurobiol.,* vol. 15, pp. 491-528, 2017.
[http://dx.doi.org/10.1007/978-3-319-57193-5_20] [PMID: 28674995]

[24]   M.N. Sonuç Karaboğa, and M.K. Sezgintürk, "Cerebrospinal fluid levels of alpha-synuclein measured using a poly-glutamic acid-modified gold nanoparticle-doped disposable neuro-biosensor system", *Analyst (Lond.),* vol. 144, no. 2, pp. 611-621, 2019.
[http://dx.doi.org/10.1039/C8AN01279B] [PMID: 30457584]

[25]   C. Erkmen, G. Aydoğdu Tığ, G. Marrazza, and B. Uslu, "Design strategies, current applications and future perspective of aptasensors for neurological disease biomarkers", *Trends Analyt. Chem.,* vol. 154, p. 116675, 2022.
[http://dx.doi.org/10.1016/j.trac.2022.116675]

[26]   S. Pineda, Z. J. Han, and K. Ostrikov, "Plasma-enabled carbon nanostructures for early diagnosis of neurodegenerative diseases", *Materials.,* vol. 7, no. 7, pp. 4896-929, 2014.
[http://dx.doi.org/10.3390/ma7074896]

[27]   H. Adam, S.C.B. Gopinath, M.K.M. Arshad, N.A. Parmin, and U. Hashim, "Distinguishing normal and aggregated alpha-synuclein interaction on gold nanorod incorporated zinc oxide nanocomposite by electrochemical technique", *Int. J. Biol. Macromol.,* vol. 171, no. 171, pp. 217-224, 2021.
[http://dx.doi.org/10.1016/j.ijbiomac.2021.01.014] [PMID: 33418041]

[28]   M. Mir, S. Palma-Florez, A. Lagunas, M.J. López-Martínez, and J. Samitier, "Biosensors integration in blood–brain barrier-on-a-chip: emerging platform for monitoring neurodegenerative diseases", *ACS Sens.,* vol. 7, no. 5, pp. 1237-1247, 2022.
[http://dx.doi.org/10.1021/acssensors.2c00333] [PMID: 35559649]

[29]   J. Kim, A. S. Campbell, B. E. de Ávila, and J. Wang, "Wearable biosensors for healthcare monitoring", *Nature biotechnology.,* vol. 2, no. 4, pp. 389-406, 2019.
[http://dx.doi.org/10.1038/s41587-019-0045-y]

[30]   H. Park, W. Park, and C.H. Lee, "Electrochemically active materials and wearable biosensors for the *in situ* analysis of body fluids for human healthcare", *NPG Asia Materials.,* vol. 13, no. 1, pp. 1-14, 2021.
[http://dx.doi.org/10.1038/s41427-020-00280-x]

[31]   J.R. Sempionatto, M. Lin, L. Yin, E. De la paz, K. Pei, T. Sonsa-ard, A.N. de Loyola Silva, A.A. Khorshed, F. Zhang, N. Tostado, S. Xu, and J. Wang, "An epidermal patch for the simultaneous monitoring of haemodynamic and metabolic biomarkers", *Nat. Biomed. Eng.,* vol. 5, no. 7, pp. 737-748, 2021.
[http://dx.doi.org/10.1038/s41551-021-00685-1] [PMID: 33589782]

[32]   F. Asci, G. Vivacqua, A. Zampogna, V. D'Onofrio, A. Mazzeo, and A. Suppa, "Wearable electrochemical sensors in Parkinson's disease", *Sensors (Basel)*, vol. 22, no. 3, p. 951, 2022.
[http://dx.doi.org/10.3390/s22030951] [PMID: 35161694]

[33]   L.C. Kourtis, O.B. Regele, J.M. Wright, and G.B. Jones, "Digital biomarkers for Alzheimer's disease: the mobile/wearable devices opportunity", *NPJ Digit. Med.*, vol. 2, no. 1, p. 9, 2019.
[http://dx.doi.org/10.1038/s41746-019-0084-2] [PMID: 31119198]

[34]   A.C. Cote, R.J. Phelps, N.S. Kabiri, J.S. Bhangu, and K.K. Thomas, "Evaluation of wearable technology in dementia: A systematic review and meta-analysis", *Front. Med. (Lausanne)*, vol. 7, p. 501104, 2021.
[http://dx.doi.org/10.3389/fmed.2020.501104] [PMID: 33505979]

[35]   N. Saif, P. Yan, K. Niotis, O. Scheyer, A. Rahman, M. Berkowitz, R. Krikorian, H. Hristov, G. Sadek, S. Bellara, and R.S. Isaacson, "Feasibility of using a wearable biosensor device in patients at risk for Alzheimer's disease dementia", *J. Prev. Alzheimers Dis.*, vol. 7, no. 2, pp. 104-111, 2020.
[PMID: 32236399]

[36]   A. Iaboni, S. Spasojevic, K. Newman, L. Schindel Martin, A. Wang, B. Ye, A. Mihailidis, and S.S. Khan, "Wearable multimodal sensors for the detection of behavioral and psychological symptoms of dementia using personalized machine learning models", *Alzheimers Dement. (Amst.)*, vol. 14, no. 1, p. e12305, 2022.
[http://dx.doi.org/10.1002/dad2.12305] [PMID: 35496371]

[37]   H. Altug, S.H. Oh, S.A. Maier, and J. Homola, "Advances and applications of nanophotonic biosensors", *Nat. Nanotechnol.*, vol. 17, no. 1, pp. 5-16, 2022.
[http://dx.doi.org/10.1038/s41565-021-01045-5] [PMID: 35046571]

[38]   C. Justino, A. Duarte, and T. Rocha-Santos, "Recent progress in biosensors for environmental monitoring: A review", *Sensors (Basel)*, vol. 17, no. 12, p. 2918, 2017.
[http://dx.doi.org/10.3390/s17122918] [PMID: 29244756]

[39]   S. Rodriguezmozaz, M. Alda, M. Marco, and D. Barceló, "Biosensors for environmental monitoringA global perspective", *Talanta*, vol. 65, no. 2, pp. 291-297, 2005.
[http://dx.doi.org/10.1016/S0039-9140(04)00381-9] [PMID: 18969798]

[40]   A. Amine, H. Mohammadi, I. Bourais, and G. Palleschi, "Enzyme inhibition-based biosensors for food safety and environmental monitoring", *Biosens. Bioelectron.*, vol. 21, no. 8, pp. 1405-1423, 2006.
[http://dx.doi.org/10.1016/j.bios.2005.07.012] [PMID: 16125923]

[41]   M. Campàs, B. Prieto-Simón, and J.L. Marty, "Biosensors to detect marine toxins: Assessing seafood safety", *Talanta*, vol. 72, no. 3, pp. 884-895, 2007.
[http://dx.doi.org/10.1016/j.talanta.2006.12.036] [PMID: 19071702]

[42]   H. Vaisocherová, A.D. Taylor, S. Jiang, K. Hegnerová, M. Vala, J. Homola, B.J. Yakes, J. Deeds, and S. DeGrasse, "Surface plasmon resonance biosensor for determination of tetrodotoxin: prevalidation study", *J. AOAC Int.*, vol. 94, no. 2, pp. 596-604, 2011.
[http://dx.doi.org/10.1093/jaoac/94.2.596] [PMID: 21563695]

[43]   R. Garjonyte, Y. Yigzaw, R. Meskys, A. Malinauskas, and L. Gorton, "Prussian Blue- and lactate oxidase-based amperometric biosensor for lactic acid", *Sens. Actuators B Chem.*, vol. 79, no. 1, pp. 33-38, 2001.
[http://dx.doi.org/10.1016/S0925-4005(01)00845-0]

[44]   N. Nikolaus, and B. Strehlitz, "Amperometric lactate biosensors and their application in (sports) medicine, for life quality and wellbeing", *Mikrochim. Acta*, vol. 160, no. 1-2, pp. 15-55, 2008.
[http://dx.doi.org/10.1007/s00604-007-0834-8]

[45]   F. Davis, and S.P.J. Higson, "Carrier systems and biosensors for biomedical applications", *InTissue Engin. Using Cer. and Poly.*, no. Jan, pp. 270-302, 2014.
[http://dx.doi.org/10.1533/9780857097163.2.270]

[46]   B.V. Ribeiro, T.A.R. Cordeiro, G.R. Oliveira e Freitas, L.F. Ferreira, and D.L. Franco, "Biosensors for the detection of respiratory viruses: A review", *Talanta Open,* vol. 2, p. 100007, 2020.
[http://dx.doi.org/10.1016/j.talo.2020.100007] [PMID: 34913046]

[47]   L.C. Brazaca, C.B. Bramorski, J. Cancino-Bernardi, S. da Silveira Cruz-Machado, R.P. Markus, B.C. Janegitz, and V. Zucolotto, "An antibody-based platform for melatonin quantification", *Colloids Surf. B Biointerfaces,* vol. 171, pp. 94-100, 2018.
[http://dx.doi.org/10.1016/j.colsurfb.2018.07.006] [PMID: 30015143]

[48]   L.C. Brazaca, B.C. Janegitz, J. Cancino-Bernardi, and V. Zucolotto, "Transmembrane protein-based electrochemical biosensor for adiponectin hormone quantification", *ChemElectroChem,* vol. 3, no. 6, pp. 1006-1011, 2016.
[http://dx.doi.org/10.1002/celc.201600099]

[49]   L.C. Clark Jr, and C. Lyons, "Electrode systems for continuous monitoring in cardiovascular surgery", *Ann. N. Y. Acad. Sci.,* vol. 102, no. 1, pp. 29-45, 1962.
[http://dx.doi.org/10.1111/j.1749-6632.1962.tb13623.x] [PMID: 14021529]

[50]   P. Palladino, A.M. Aura, and G. Spoto, "Surface plasmon resonance for the label-free detection of Alzheimer's β-amyloid peptide aggregation", *Analy. and bioanaly. chem.,* vol. 408, no. 3, pp. 849-854, 2016.
[http://dx.doi.org/10.1007/s00216-015-9172-6]

[51]   F.T.C. Moreira, M.G.F. Sale, and M. Di Lorenzo, "Towards timely Alzheimer diagnosis: A self-powered amperometric biosensor for the neurotransmitter acetylcholine", *Biosens. Bioelectron.,* vol. 87, pp. 607-614, 2017.
[http://dx.doi.org/10.1016/j.bios.2016.08.104] [PMID: 27616286]

[52]   J.V. Rushworth, A. Ahmed, H.H. Griffiths, N.M. Pollock, N.M. Hooper, and P.A. Millner, "A label-free electrical impedimetric biosensor for the specific detection of Alzheimer's amyloid-beta oligomers", *Biosens. Bioelectron.,* vol. 56, pp. 83-90, 2014.
[http://dx.doi.org/10.1016/j.bios.2013.12.036] [PMID: 24480125]

[53]   W. Ma, L.X. Qin, F.T. Liu, Z. Gu, J. Wang, Z.G. Pan, T.D. James, and Y.T. Long, "Ubiquinone-quantum dot bioconjugates for *in vitro* and intracellular complex I sensing", *Sci. Rep.,* vol. 3, no. 1, p. 1537, 2013.
[http://dx.doi.org/10.1038/srep01537] [PMID: 23524384]

[54]   S.R. Ankireddy, and J. Kim, "Selective detection of dopamine in the presence of ascorbic acid *via* fluorescence quenching of InP/ZnS quantum dots", *Int. J. Nanomedicine,* vol. 10, no. Spec Iss, pp. 113-119, 2015.
[PMID: 26347250]

[55]   L. Parnetti, D. Chiasserini, G. Bellomo, D. Giannandrea, C. De Carlo, M.M. Qureshi, M.T. Ardah, S. Varghese, L. Bonanni, B. Borroni, N. Tambasco, P. Eusebi, A. Rossi, M. Onofrj, A. Padovani, P. Calabresi, and O. El-Agnaf, "Cerebrospinal fluid Tau/α-synuclein ratio in Parkinson's disease and degenerative dementias", *Mov. Disord.,* vol. 26, no. 8, pp. 1428-1435, 2011.
[http://dx.doi.org/10.1002/mds.23670] [PMID: 21469206]

CHAPTER 5

Biosensors for Protein Bio-Sensing and Detection of Bacteria and Viruses

Himani Yadav[1], Bhaskar Sharma[2], Priti Giri[1], Avanish Kumar Shrivastav[3], Vivek Kumar Chaturvedi[4], Prem L. Uniyal[1] and Ravi Kumar Goswami[5,*]

[1] *Department of Botany, University of Delhi, Delhi-110007, India*

[2] *Neurobiology Laboratory, Department of Anatomy, All India Institute of Medical Sciences, New Delhi, 110029, India*

[3] *Department of Biotechnology, Delhi Technological University, Delhi-110042, India*

[4] *Department of Gastroenterology, Institute of Medical Sciences, Banaras Hindu University, Varanasi, Uttar Pradesh, India*

[5] *Department of Zoology, Hindu College, University of Delhi, New Delhi-110007, India*

Abstract: The concept of a biosensor was first proposed in 1962 by Clark and Lyons, who developed an oxidase enzyme electrode for the detection of glucose. Since then, the development of nanotechnology has prompted biosensors to evolve and become more specialized for a variety of applications. Currently, at the forefront of science, bio-sensing applications combined with nanotechnology have implications for multiple fields, including medicine, biology, environment, drug delivery, and food safety. In recent decades, bacterial and viral diseases have seriously threatened human safety. Prioritizing the rapid detection of outbreaks, which pose a major threat to the healthcare system and could have a catastrophic socioeconomic impact, will help stop them. Scientists are conducting extensive research to develop sensitive diagnostic techniques and effective medicines.

Keywords: Biosensors, Bacteria, Diagnosis, Nanotechnology, Viruses.

INTRODUCTION

The idea of a biosensor was initially brought up by Clark and Lyons in 1962 when they created an oxidase enzyme electrode for the detection of glucose [1]. Since then, the growth of nanotechnology has encouraged the evolution and specialisation of biosensors for various applications [2]. Nanotechnology is currently at the cutting edge of science, and its combination with bio-sensing

---

[*] **Corresponding author Ravi Kumar Goswami:** Department of Zoology, Hindu College, University of Delhi, New Delhi-110007, India; Email: ravigoswamigen@gmail.com

Vivek K. Chaturvedi, Dawesh P. Yadav and Mohan P. Singh (Eds.)

applications affects a variety of sectors, including medicine, biology, the environment, drug delivery, and food safety [3].

However, considering the importance of bacterial and viral diseases in human health today, the detection of pathogens has emerged as one of the most important goals for these devices. The reverse transcription polymerase chain reaction (RT-PCR), the gold standard for pathogen identification, is widely used to detect viruses and bacteria. As part of standard detection methods, these pathogens typically must be isolated, cultured, and subjected to biochemical testing. In addition, serological tests such as ELISA are used to detect antibodies and immunoglobulins, which are essential for identification [4]. However, some of these methods are labour intensive and often take a long time to produce results.

Although there are numerous techniques for locating viral particles, their practical application is limited by several difficulties. These restrictions consist of:

1. Less precise and sensitive

2. The requirement for sample cleaning and preparation

3. It takes time

4. Increased instrument, accessory, and upkeep costs

5. Widespread accessibility

6. The instruments' intricate operation

7. The need for highly qualified technical staff

8. Insufficient for quick, on-site analysis

In light of viruses' adaptability and potential reproduction sites, there is a need for more advanced, more efficient methods for the rapid detection of viral analytes. These approaches must ensure greater accuracy, portability, ease of use, and widespread accessibility for testing the general population. As a result, novel strategies based on nanotechnological developments have arisen as viable and simple solutions for quickly and effectively identifying infections. On the one hand, nanoparticles (NPs) have proven to have exceptional anti-pathogen characteristics and have been exploited to create novel devices and solutions that help with this public health concern [5]. Since zoonosis is a real concern, the focus extends beyond human illnesses to include those affecting animals. Gold nanoparticles (AuNPs) and quantum dots (QDs) were used to create an optical biosensor for the detection of the porcine reproductive and respiratory disease virus [6]. On the other hand, there is a growing interest among scientists worldwide in using DNA biosensors or sequence-specific DNA detectors for clinical studies. A combination DNA-based piezoelectric biosensor was created for the simultaneous detection and genotyping of high-risk human papillomavirus (HPV) strains [7].

These gadgets are useful *(e.g.,* they allow point-of-care (POC) testing using a nano-biosensor built into a smartphone), quick, and are regarded as novel technologies that offer an alternative solution to the drawbacks of conventional detection methods that have been stated [8]. These technologies have been used to investigate bacteria like *Escherichia coli* and *Salmonella spp.* as well as viruses like the Ebola virus, human immunodeficiency virus (HIV), and more recently, acute respiratory syndrome coronavirus 2 (SARS-CoV-2) that affect human health [9]. Biosensors are analyte detection devices that combine a biological component with a physicochemical detector known as a biosensor [10]. The biosensor's design and intended use affect how an analyte is detected. With the addition of simple accessories, several everyday items, like cell phones, can be utilised as biosensors, where they built a non-invasive smartphone-based urea biosensor using saliva as a sample [11]. This enables quick and affordable preliminary detection [12]. Typically, biosensors identify substances linked to disease, including nucleic acids, proteins, and cells. The physiologically sensitive element, the detector element, and the reader device are their three main constituents, which allow for this. The biomolecules are identified using nucleic acids, organelles, antibodies, enzymes, and microbes. Researchers also need to determine the prerequisites for obtaining a functional gadget per the desired usage.

Therefore, multidisciplinary research is essential to choosing the right material, transducing technology, and biological components before putting the biosensor together [13]. Biosensors are used in the clinical setting to find biomolecules linked to disease. These tools are capable of detecting biochemical disease indicators in bodily fluids, including saliva, blood, or urine. A non-invasive glucose testing technique is based on a dispensed saliva nano-biosensor to increase patient compliance, lessen complications, and lower expenses associated with managing diabetes. In comparison to the UV spectrophotometer, they achieved exceptional accuracy outcomes in the clinical testing. The disposable gadget can therefore be suggested as an option for tracking salivary glucose in real-time [14]. Numerous more clinical diagnostic uses for biosensors exist, including cholesterol, cardiovascular disease markers, cancer or tumour biomarkers, allergic reactions, and infections with pathogenic bacteria, viruses, and fungi. In addition, biosensors can be used to detect bacteria and viruses in water and food, which are potential disease-causing agents. For the quick *in-situ* detection of *E. coli,* a low-cost, portable microfluidic chemo resistive biosensor is based on monolayer graphene, AuNPs, and streptavidin-antibody system. In this instance, the biosensor's surface is used to catch the bacteria, and electric readouts are used to detect them [15]. Using a self-assembled monolayer, disclose the creation of a glutathione-S-transferase tag for white spot binding protein (GST-WBP) anchored onto a gold electrode. The binding between the White Spot Syndrome Virus (WSSV) and the immobilised GST-WBP enables this biosensor to detect WSSV in shrimp pond water [16].

OPERATING PRINCIPLES

Three operational elements combine to form a biosensor. These devices feature sensing components, also referred to as bioreceptors, that mimic *in vivo* molecular recognition processes. Numerous different sensing components exist, including cells, microorganisms, cell receptors, antibodies, enzymes, and nucleic acids [17]. Depending on the type of biosensor, these biologically sensitive components identify the analyte and engage with it. Nucleic acid sequences from bacteria or viruses serve as the foundation for one of the primary biorecognition techniques. To detect *Vibrio cholerae*, a DNA bioelectrode that is stable for at least 15 weeks when stored at 4 °C. The biosensor was made of sol-gel-produced nanostructured zirconium film and a 24-mer single-stranded DNA probe based on the O1 gene [18]. Numerous clinical diagnostics uses for biosensors exist, including cholesterol, cardiovascular disease markers, cancer or tumour biomarkers, allergic reactions, and infections with pathogenic bacteria, viruses, and fungi. In addition, biosensors can be used to detect bacteria and viruses in water and food, which are potential disease-causing agents. For the quick *in-situ* detection of *E. coli*, a low-cost, portable microfluidic chemo resistive biosensor based on monolayer graphene, AuNPs, and streptavidin-antibody system. In this instance, the biosensor's surface is used to catch the bacteria, and electric readouts are used to detect them [15]. By using a self-assembled monolayer, disclose the creation of a glutathione-S-transferase tag for white spot binding protein (GST-WBP) fixed onto a gold electrode. Due to the interaction between the immobilised GST-WBP and the white spot syndrome virus (WSSV), this biosensor can detect WSSV in shrimp pond water [19]. Depending on the type of biosensor, these biologically sensitive components interact with the analyte after recognising it. Based on bacterial or viral nucleic acid sequences, one of the primary biorecognition techniques is used. A DNA bioelectrode for *Vibrio cholerae* detection was created and is stable for at least 15 weeks when stored at 4 °C. The biosensor used a 24-mer single-stranded DNA probe based on the O1 gene mounted on a sol-gel produced nanostructured zirconium oxide sheet [18].

The transducer, also known as a detector, is the second component. It functions by detecting the signal associated with a physicochemical change brought on by the interaction of the bioreceptor and the analyte. The signal is changed into a different one that may be measured and analysed. The reading device is the biosensor's final component. Typically, a display is used, and both hardware and software are used to produce the results. The performance of a biosensor is determined by a few key factors. First and foremost, selectivity is the ability of a bioreceptor to identify a particular bio-entity in a sample made up of other components. This is most likely the primary characteristic that determines the required bioreceptor. Second, repeatability is the capacity for an experimental setup to yield the same result when repeated numerous times. Signals with high reproducibility and robustness are available. Thirdly, stability is the ability to withstand outside disturbances that could compromise the device's accuracy and precision. Fourth, sensitivity, commonly referred to as the limit of detection (LOD), is the lowest concentration of the analyte that a biosensor can detect. It is necessary to identify the analyte in samples with low concentrations for clinical applications. Last but not least, linearity evaluates how accurate measurements are in response to the smallest concentration variation that might change the output *(i.e.,* resolution) and within the linear range of analyte concentrations [20].

TYPES OF BIOSENSORS

According to how they convert signals into optical, electrochemical, and piezoelectric devices, biosensors can be categorised [21]. Optic fibres are used as the transduction medium by optical biosensors, which measure photons for their analysis [22]. This kind of biosensor can detect analytes using a variety of optical sensing techniques, such as absorption, colourimetry, fluorescence, or luminescence. Compared to electrochemical and piezoelectric biosensors, this type of biosensor exhibits less noise and is immune to electromagnetic interference. A chromatic biosensor for speedy bacterial detection was created, based on composites of non-woven fibres made of polyvinyl butyrate and polydiacetylene. The tool has the potential to provide early warnings of probable infections brought on by *Staphylococcus aureus, Micrococcus luteus*, and *E. coli* [23]. A fluorescent supramolecular biosensor for bacterial detection in a different investigation. Fluorescence emission brought on by conformational changes brought on by the interaction of these pathogens allows for the selective detection of *E. coli* over other microbes [24]. Viral particles adhering to a microsphere optical resonator caused the resonance to shift to longer wavelengths, which was tested as an optical biosensor for single virus detection in viral analysis. Furthermore, the advancement of immunoassays has tremendously benefited Surface Plasmon Resonance (SPR), an optical method. When a metallic thin film is deposited on a dielectric waveguide, this type of resonance occurs, and it is determined by using the intensity data from the reflection of p-polarized light (*i.e.*, along the plane of incidence). Recently, an electrochemical biosensor with non-cytotoxic silica nanoparticles was used to detect *E. coli* with high sensitivity and specificity. By using cyclic voltammetry readings, the electrochemical immune biosensor can identify bacteria in five minutes. It also has promise as a tool that may be modified to target several different microbes [25]. In another study, an electrochemical biosensor for the detection of human norovirus was developed, consisting of eight novel peptides individually loaded into a gold electrode. The peptides had a high binding affinity for the viruses, and an increase in virus concentration was used to explain a decrease in current signals [26]. Piezoelectric biosensors are the last but most significant. A material's capacity to produce a voltage when subjected to mechanical stress is known as piezoelectricity. These biosensors have crystals that vibrate when an electric field is applied. In addition, several substances react to interactions with other molecules by vibrating at distinctive resonance frequencies. The working basis of transduction in this kind of biosensor is thought to be the relationship between the resonant frequency variations and the mass from the molecules adsorbed or desorbed from the crystal's surface. Vibration thus offers knowledge about the process being measured [27].

Role of Biosensors in the Detection of a Variety of Pathogens

Escherichia Coli

Escherichia coli is the most common type of bacteria linked to waterborne and foodborne illnesses. The requirement to cope with episodes prompted researchers to develop logical tools to recognise elements of the bacterial surface [28]. The methods included the use of antibodies and aptamers to achieve the immobilisation of nanoparticles [29 - 31]. Numerous systems have been used, such as Polymerase Chain Reaction (PCR) [32, 33], circle interceded isothermal enhancement (LAMP) [34], and surface plasmon reverberation (SPR) [35 - 39]. Other coordinated microfluidic frameworks comprised the positioning approach, elution arrangement, DNA extraction channel, and washing arrangement [40, 41]. A very intriguing way suggested using a cell phone for visual augmentation and location or evaluation [42]. The method was successfully tested on a variety of grids, including those with drinking water, milk, blood, and spinach. A different researcher developed an electrical model that can assess the number of immobilised antibodies and is biomolecule-viable [43]. The location and specificity of *E. coli* 16S ribosomal RNA in meat testing served as the basis for an approach [44]. Recent mechanical developments have allowed researchers to develop a radial microfluidic programmed remote endpoint locating framework coordinated with LAMP that is capable of 30 simultaneous responses [45].

Similar studies have examined microfluidic devices using the LAMP methodology [41]. A combination of dendrimers and aptamers as acknowledgement components for *E. coli* location using a microfluidic architecture. Due to the numerous restriction areas on the specified cells, this technique improved attentiveness [10].

Salmonella Species

One of the main causes of foodborne illness is *Salmonella* spp. The culture media strategy is the highest level of testing methodology, but it takes 2–5 days to confirm a result. Numerous works have focused on foods like meat, dairy, vegetables, beverages, and society. For example, quantum dabs [46], appealing dots [29] and quartz gems are nanoparticles that are utilised to think of microorganisms [47]. Dots produced with a cathode or antibodies and functionalized with certain aptamers were also used to achieve restriction. LAMP [34], PCR [32, 48], impedimetric sensors [49, 50], acoustic wave sensors [16], and SPR were some of the techniques used [35]. It is possible to develop a coordinated microfluidic sensor that provides precise results using image analysis done with a PDA application [51]. Despite ongoing research in the field of biosensors, there is still a need to evaluate and enhance biosensor devices, cut down on process time, lower costs, and work on geographical constraints. Various biosensor frameworks have been taken into account recently to address the aforementioned problems. To locate monoplex and multiplex *S. typhimurium* and *Vibrio parahaemolyticus*, a connected framework was developed with a coordinated turning microfluidic framework suitable for DNA extraction, LAMP, and a colorimetric parallel stream strip [52]. A radial microfluidic stage on a single plate was what kept the cycle going. An exciting study combines the PCR and SPR techniques in a single device with the ability to perform amplicon localization without using a fluorophore. This method enabled accurate estimation using the SPR fibre sensor component in addition to improving the DNA of *S. typhimurium*. The SPR fibre sensor offered advantages for analytical applications and was reusable. For the location of *S. enteritidis,* a microbe evaluation framework using a basic film and microfluidics was introduced [53, 54]. The device was designed to simultaneously improve quality, mix arrangements, and locate electrochemically using polyester and polyimide films de-stretched onto a polycarbonate housing chip. Another method of locating used antibodies that were focused on appealing dabs for restricting [55]. Direct PCR was then conducted without the need for DNA *ex vivo* and the associated item loss. Various instances, including a vegetable plate of mixed greens, an egg yolk, an egg white, a complete egg, and minced pork meat, demonstrated a location constraint of 2 CFU/mL. The ability to carry out the necessary systems for *Salmonella, E. coli* and *S. aureus* location in food testing was provided by a lab-on-a-chip approach that focused on microdevices. The LAMP method was used to completely extract and enrich DNA using a paper-inserted device. Within 75 minutes, colorimetric adjustments produced a response. A web-based technique for the detection of *Salmonella* was promoted [56]. To increase the explicitness, the device concentrates microorganisms with harmlessly alluring nanoparticles. The safe fluorescent microspheres, which allowed fluorescence outflow, were named after the appealing microorganisms. The fluorescence could be evaluated with a PDA application. For the location of *Salmonella* with high sensitivity, a quick screening method was suggested. Polystyrene microspheres (PSs) were modified with hostility to *Salmonella* polyclonal antibodies and catalases, and attractive nanoparticles were modified against *Salmonella* monoclonal antibodies to bind bacterial cells. Through a polymer tube, the alluring nanoparticles were delivered to the PSs, where the electrical voltage change was monitored [56].

Listeria Monocytogenes

One of the main microorganisms found in food is *L. monocytogenes*. Its occurrence in fish and beverage matrices and societies has been studied. Studies on its location have mostly focused on optical and impedimetric techniques. These have been built on antibodies, nucleic acids, modified attractive nanoparticles, and platinum microelectrodes [32, 57, 58]. The localization methods were impedimetric estimate, PCR, and LAMP [59]. Multiplex PCR has also been evaluated [60]. A colorimetric assay for *L. monocytogenes* localisation using alluring nanoparticles was described in another review. To tie with Listeria protease and produce a variety of alterations in sullied arrangements containing gold nanoparticles and a few chemicals, a D-amino corrosive was used as a substrate. The method could identify the bacteria at 2.17 102 CFU/mL. An insightful evaluation used an optical biosensor to identify the high quality of *L. monocytogenes* in raw and frozen fish using a lab-on-paper chip [61]. This season's virus fluorescence signal was provided by SYBR colour, and the evaluation was synchronised with LAMP on a single chip stage. With a location threshold of 100 copies of the objective quality, the test produced good responsiveness (100%) results. A radial microfluidic device was used as the final coordinated programmed sub-atomic sensor to find Listeria among three unrelated bacteria [41]. Positions for test stacking, a washing arrangement, an elution arrangement, and a LAMP blend were all included in the bad habit. The reagents are provided when the device rotates. The system was completed in 65 minutes and allowed for the simultaneous handling of 18 instances and two controls. In milk tests, the limit of location for *Listeria* was 103 CFU/mL.

Vibrio Species

Every year, cholera accounts for 21,000 to 143,000 fatalities and 1.3 to 4.0 million cases worldwide [62]. The techniques have been used on bacterial societies, dairy products, beef, fish, and seafood. Preconcentration is carried out using a bacteriophage paired with luciferase, a bioluminescent framework, and particles identified with antibodies [63]. The identification of *Vibrio* spp. was often visualised by optical detection, LAMP, and glow [41]. The development of a microfluidic device that enabled control with a crisscross-shaped microchannel and RPM. To locate *Vibrio parahaemolyticus* in a fixed fluid stage lab-on-a-chip, a charming technique was used. Using this technique, it was possible to capture particles that had been immunised as well as named particles that had been functionalized with horseradish peroxidase and a body adversary. The device included a locating chamber, a washing chamber, and an example chamber that were all connected by two channels. The two types of particles were combined in the example chamber with the example, and the resulting agglomeration was then transferred to the location chamber with a chromogenic substrate solution. *V. parahaemolyticus* was identified in contaminated clam testing with a location restriction of 10 cells. To transfer results to a PDA, a radial microfluidic device with Bluetooth remote technology was developed. The LAMP confirmed the location [45]. In a 60-minute butt-centric analysis, the microdevice could carry out 30 responses simultaneously for several species and strains of *V. cholerae*. A combined system made up of a calorimetric sidelong stream strip, a LAMP, and a coordinated rotating microfluidic framework appropriate for DNA extraction. *S. typhimurium* and *V. parahaemolyticus* were located in front of a centrifugal microfluidic stage on a single plate for multiplex measurement. 50 CFU/test served as the detection limit [52].

Streptococcus Species

Streptococcus spp. is a major cause of human and veterinary grimness and death worldwide, among other bacteria. *Streptococcus* screening for milk and water tests using biosensors has been suggested. With enticingly named explicit antibodies and peptides, the explicitness was increased [64]. SPR an impedance cluster [65], a cytometer, and an electrical sensor were present at the site [35, 66]. Additionally, multiplex testing has been tried [65]. Streptococcus agalactiae caused ox-like mastitis was discovered during a device survey in a rural location. Using a lab-on-a-chip magnetoresistive cytometer, milk tests were selected and combined with a solution that combined specific antibodies and attracted nanoparticles to achieve quick and precise localization [64]. This immunological localization method enabled the identification of 100 CFU/mL. Khan and collaborators made an effort to improve the location and reaction time constraints for the detection of bacterial strains inside a single sensor chip. For continually catching both Gram-positive and Gram-negative microorganisms, they presented an electrically open, thermally responsive sensor conjugated with a composite polymer with disappearing Au electrodes. Milk, lake water, tap water, and autoclave water were all included as examples in the grids. The maximum cell density for *E. coli, B. subtilis*, and *S. mutans* was 5 cells/mL [31].

OTHER BACTERIA

Campylobacteriosis is one of the most well-known water and foodborne illnesses in the European Union. According to explicit phage biorecognition studies mounted on the sensor surface, a microresonator demonstrated assurance for *Campylobacter jejuni* localisation. It made use of high reverberation frequencies possible. To support the specificity of location and to weed out signals from unrelated microbes, various bacterial cells were tested. With a detection limit of 91011 mol/L, the first electrochemical genosensor was utilised to detect *Campylobacter* spp. in poultry flesh [67]. This genosensor was based on slim film gold electrodes preserved on a Cyclo Olefin Polymer substrate. The anode surface requires electrochemical activity, which focuses on the device's excellent acid sensitivity. The development of a biosensor with antimicrobial peptides with species-explicit concentrating and restricting capabilities was driven by the urgent need for the location of a variety of microorganisms in the food industry (Table **1**) [65]. Through the use of cysteine gold science, *Streptococcus mutans* and *Pseudomonas aeruginosa* interacted with specific peptides immobilised on a gold surface, producing an electrical signal. The minimum focus determined during 25 minutes was 105 CFU/mL. Another study used a conductometric sensing device with a glass coating that was less than one millimetre thick to detect *Pseudomonas putida* [38]. When the target 16S rRNA was limited, the device connected oligonucleotide test polystyrene dots, which were electrically portable with a noticeable step drop in ionic flow. Compact, user-friendly biosensors were required for application in temperature control and water treatment. A synchronised aloof stream optical microfluidic device was therefore developed. For testing chemiluminescent sandwich immunoassays enhanced by gold nanoparticles, ring-shaped natural photodiodes were combined into a thin, incited stream microfluidic channel [68]. *Legionella pneumophila* 4 104 cells/mL could be found in water testing according to the method. Studies have been conducted on the identification of cyanobacteria that create protease inhibitor oligopeptides, such as cyanopeptolins, which contaminate drinking water. The biochip includes Au cathode displays and a microfluidics architecture that allows a regulated reagent stream for rapid DNA localization [69]. The findings suggested limiting the biotinylated discovery test, which uses an amperometric estimate and provides a constraint of recognition objective DNA, to the avidin and protein-modified Au nanoparticles. With a small SPR biosensor using antibodies, the difficulty typically encountered in spore location was eliminated [70]. According to the authors, this was the first major study of the differential location of *Bacillus globigii* using the aforementioned method in a mixed sample that included at least one additional *Bacillus* species. With immunizer infusion and direct catch, the locational constraint was 105 spores/mL and 107 spores/mL, respectively. One of the most expensive diseases for dairy farmers in the country is ox-like mastitis. *Staphylococcus aureus* frequently causes the condition. Milk tests were chosen, combined with a solution that included specific antibodies and appealing nanoparticles, and then evaluated using a lab-on-a-chip magnetoresistive cytometer to detect the bacterium [71]. This method of immunological detection enabled the recognition of *S. Aureus*. A fascinating study [72] described the use of a phage clone that may form a specific complex with the surface of the *S. aureus* cell. Biosensors based on bacteriophages have several advantages because of their high specificity for a specific bacteria, vitality, ease and low cost of production. For precise bacterial identification, a mica surface was used to immobilise the phage clone. Almost half of the cells were quickly recognized thanks to the physisorbed phage. After a 30 minute incubation and hydration, a coordinated biosensor framework was developed for each mitting the position of three different species. The LAMP method was used for amplification and the colorimetric response was visible to the naked eye. The device was used continuously to distinguish between the bacteria *E. coli*, *Salmonella* spp., and *S. aureus* [73].

VIRAL ILLNESSES

A coordinated programmable electrochemical immunosensor was developed to determine the presence of hepatitis infections [74]. The device had immobilised infection antibodies that were captured on an electrochemical sensor display that captured antigens from the setup. The model has been used to determine where viral proteins are present. For the location of both diseases and microorganisms, a suggestion was made [60]. Through regular fluorescence combined with an alluring burrowing intersection, measurements for the site of normal DNA extracted from Hepatitis A (HAV), Hepatitis E (HEV), *Listeria,* and *Salmonella* were developed and tested. The DNA assays made it possible to assess numerous bacteria continuously in a single measurement. Late investigations focused on the site of the norovirus, a foodborne germ that can cause erratic and persistent gastrointestinal symptoms. A concanavalin A-formed nano-organized gold cathode was used to develop an electrochemical biosensor [75]. The framework was able to detect the presence of small groups with 60 duplicates per millilitre. A novel idea was put into practice by using an aptamer and scaled-down microelectromechanical frameworks to create a small electrochemical sensor. To correlate the infection, which was determined using cyclic voltammetry with influenza florescence perception, a specific aptamer for the Norovirus was produced on a gold working cathode [76].

Table 1. Records for biosensor applications in food and water.

Microorganism	Matrix	Method	Limit of Detection (LOD)	Reference
S. enterica	Culture	Microcantilever	10^3 CFU/mL	[91]
E. coli K12	Cultured bacteria	Electrochemical biosensor	10^3 CFU/mL	[43]
E. coli	Fresh spinach	Surface plasmon resonance (SPR)	10^3 CFU/ mL	[28]
HAV, HEV	Culture	Electrochemical immunosensor array	0.5 ng/mL, 1.0 ng/mL	[74]
E. coli	Apple juice, orange juice	Magneto-elastic (ME) resonant μ-diver system (MER-μds) is proposed and prototyped	10^2 CFU/mL	[37]
S. enterica serovar Enteritidis,	Mineral water, raw cow	SPR	2.8 CFU/mL	[35]
S. pneumoniae, E. coli O157:H7	milk, frozen ground meat	–	–	–
E. coli	Culture	Electrochemical detection of microbial 16S ribosomal RNA	1 CFU/mL	[44]
E. coli	Culture	Plate count technique using arrayed microelectrodes	Not mentioned	[40]
C. jejuni	Culture	Microresonator	Not mentioned	[67]
E. coli	Culture	Optical transmission (EOT) phenomenon in plasmonic nanohole	$<10^2$ cells	[30]
S. typhimurium	Chicken breast	Microfluidic nano-biosensor	10^3 CFU/mL	[46]
V. parahaemolyticus	Aquatic sewage water, oyster	FMN: NADH oxidoreductase	10 CFU/mL	[63]
S. typhimurium	Culture	Electrical biosensor	10^3CFU/mL	[50]
S. agalactiae	Milk	Lab-on-a-chip magnetoresistive	10 CFU/μL	[66]
S. mutans, P. aeruginosa	Culture	cytometer, microfluidic Impedance biosensor	10^5 CFU/mL	[65]
L. pneumophila	Water	Optical microfluidic devices	4×10^4 cells/mL	[68]

(Table 1) cont.....

Norovirus	Lettuce	Electrochemical biosensor	6×10^1 copies/mL	[75]
Planktothrix agardhii NIVA-CYA116	Freshwater	Nanoparticles-based amperometric biosensor	6×10^{12} mol/L target DNA	[69]
Campylobacter spp.	Poultry meat	Microscale electrodes	9×10^{-11} mol/L	[92]
L. monocytogenes	Lettuce	Impedance biosensor	3×10^2 CFU/mL	[4]
E. coli O157, *E. coli*	Culture	Optical biosensor	5×10^2 cells	[33]
E. coli	Culture	Optical biosensor	$10^3 - 10^7$ CFU/ mL	[74]
E. coli	Water, milk, blood, and spinach samples	Visual detection using a smartphone	$10^1 – 10^3$ CFU/mL	[42]
Norovirus	Culture	Electrochemical	Not mentioned	[76]
S. typhimurium	Culture	Acoustic wave	10^2 BCE/sample	[16]
L. monocytogenes	Culture	Platinum interdigitated	5.39 CFU/mL	[58]
H5N1	Poultry Culture	Aptasensor fluorescence	0.4 HAU	[93]
E. coli O157:H7, *S. typhimurium, V. parahaemolyticus*		Optical biosensor- LAMP & Eriochrome Black T (EBT)	3.8×10^2 copies	[34]
Bacillus. anthracis	Culture	SPR	$10^5 – 10^7$ spores/mL	[70]
L. monocytogenes	Culture	Colorimetric sensing nanoparticles	2.17×10^2 CFU/mL	[57]
Tobacco mosaic virus	*E. coli*	Optical biosensor- LAMP	9.1 ng/mL	[94]
L. monocytogenes	Culture	Electrochemical impedance analysis	1.6×10^2 CFU/mL	[59]
S. agalactiae	Milk	Lab-on-a-chip magnetoresistive cytometer, microfluidic	10^2 CFU/mL	[64]

BIOSENSORS IN CORONAVIRUS PANDEMIC

Numerous illnesses are currently thought to have a limited ability to start pandemics in the future. They pose a serious threat to humanity because of several factors, including their rapid spread, the rapid transmission of new variations, the difficulty of developing effective and reasonable indicative methods, the lack of specific immunizations, and the lack of safe medications for treatment [77]. The most recent occurrence is COVID-19, which was declared a global pandemic on March 11, 2020, and is an infectious disease caused by the SARS-CoV-2 virus that spreads quickly from human to human. This bacterium belongs to disease caused by positive-strand RNA [78]. An early diagnosis is essential to prevent the disease from spreading like previous viral flare-ups. However, this pandemic has the characteristic that more than 30% of the confirmed cases are asymptomatic, making it more difficult to control [79]. RT-PCR is currently the most involved, reasonable, and reliable method for detecting SARS-CoV-2 contaminations. In any event, the technique is time-consuming, labour-intensive, and impractical in remote locations. Although several various approaches, such as immunological measurements, thoracic imaging, flexible X-beams, or intensification techniques, can be used for that purpose, the pandemic spread of COVID-19 necessitates the development of POC devices for rapid identification [80]. There are two different types of rapid POC biosensors available for COVID-19 detection, according to Sheridan. In any instance, a nucleic analysis is available to determine whether the infection is present in the patient's sputum, spit or nasal discharges [81]. The other type that is frequently used is the neutralizer test, which is completed by the analysis of collected blood tests five days after the underlying disease, which is the time when the safe reaction causes IgM and IgG to develop as a result of the infection's presence. Modern science has developed several suitable POC biosensors for the opportunistic detection of SARS-CoV-2 IgM and IgG antibodies using tests requiring as little as 10 L of human serum, whole blood, or finger prick, with results arriving in only 10-15 min [82]. These rapid serological tests frequently use paper-based biosensors that perform colorimetric sidelong stream immunoassays. This method ties the comparative host antibodies, which move across a cement cushion, to the SARS-CoV-2 explicit antigens, which are frequently designated with gold. IgM antibodies that are protective against SARS-CoV-2 interact with antibodies that are specifically hostile to IgM on the M line, whereas IgG antibodies cooperate with antibodies that are specifically hostile to IgG on the G line. If the example contains SARS-CoV-2 antibodies, then M or G lines may afterwards appear. In any case, only the control line (C) will be displayed. Although the use of serological tests to detect SARS-CoV-2 is still under discussion, these tests are expected to be crucial tools to implement or end the planned lockdowns [83]. For the localization of SARS-CoV-2, other research groups have developed Lab-on-a-Chip-based biosensors [84]. By incorporating microfluidic components into a biosensor, this innovation gets beyond the need for specialised training, allowing for increased production and lower research costs [85]. POC-popularized equipment like ID NOW®, Filmarray®, GeneXpert®, and RTisochip® are significantly contributing to this epidemic in light of this microfluidic breakthrough [86]. In addition, cell-based biosensors have improved COVID-19 analysis. In light of film-designed mammalian cells that exhibit the human illusory spike S1 immunizer, the device can selectively identify the SARS-CoV-2 S1 spike protein, which limits the protein to layer-bound antibodies and alters the cell's bioelectric characteristics as measured by the Bioelectric Recognition Assay. The reaction time is about three minutes, and the

LOD is 1 fg/mL. The biosensor also features a small read-out device that may be operated by a smartphone [87].

Furthermore, nano-biosensors have demonstrated a remarkable potential to contribute to the fight against COVID-19 by providing complete knowledge for developing ultrasensitive, useful, and quick discovery tools for large-scale manufacture. High-level materials serve as the foundation for nano-empowered or coordinated miniature and nano bio-sensing framework advancements that can detect infections before they occur and even exhibit excellent restricting properties that allow them to render the microbe inert or eradicate it when an external upgrade is applied. POC biosensors based on graphene and carbon have been developed by many research groups [88]. It is anticipated that graphene will be crucial to the effort to combat COVID-19. Since its awareness and selectivity may be improved by changing its crossover structure (for example, immunizer-produced graphene sheets), which allows adjustment of its optical and electrical highlights, this inexpensive material can be used for infection recognition. Photoluminescence, colorimetric, and SPR biosensors are a few graphene-based sensors that can be researched for SARS-CoV-2 location. The information was used by website optimization *et al.* to create a biosensor based on field-impact semiconductors (FET) that can detect SARS-CoV-2. In this case, a specific immunizer against the infection spike protein was applied to graphene sheets from the FET and successfully identified at centralizations of 1 fg/mL in a phosphate-supported saline medium. Additionally, the device could identify the infection in clinical cases by showing a LOD of 2.42 102 duplicates/mL. A possible immunological demonstration treatment for the illness is the produced biosensor, according to experts [89]. Additionally, nano-biosensors have demonstrated a remarkable potential to contribute to the fight against COVID-19 by providing extensive experience in developing ultrasensitive, useful, and quick identification tools for large-scale manufacture. High-level materials serve as the foundation for nano-empowered or nano-incorporated miniature and nano bio-sensing framework innovations that can detect infections before they occur and even exhibit excellent restricting properties allowing them to render the microbe inert or eradicate it upon the application of an external advancement. POC biosensors based on graphene and carbon have been developed by many exploring groups. It is anticipated that graphene will be crucial to the effort to combat COVID-19. Since its awareness and selectivity may be improved by changing its hybrid architecture (for example, neutralizer-formed graphene sheets), which enables fine-tuning of its optical and electrical features, this inexpensive material can be used for infection discovery. Photoluminescence, colorimetric and SPR biosensors are a few graphene-based sensors that can be researched for SARS-CoV-2 identification [90].

CONCLUSION

Viral and bacterial infections have posed significant threats to human security in recent decades. To stop an outbreak that has a serious risk of disrupting the healthcare system and catastrophic socioeconomic impact, its rapid detection must be a priority. Scientists conduct intensive research to develop sensitive diagnostic methods and effective drugs. Developing a POC device to quickly diagnose diseases like COVID-19 will enable life-saving decisions to be accelerated and positive patients to be isolated early. For many viruses and bacteria, there are no vaccines or pharmaceutical treatments. By sensing pertinent factors that may be associated with infectious processes, biosensors and nano-biosensors are potent measurement tools that can make the detection of significant clinical bacteria and viruses simple, quick, and effective.

ACKNOWLEDGEMENTS

V.K.C. gratefully acknowledges the Department of Health Research (DHR), Govt. of India, for support through the Young Scientist Fellowship Grant R.12014/56/2022-HR.

REFERENCES

[1]   L.C. Clark Jr, and C. Lyons, "Electrode systems for continuous monitoring in cardiovascular surgery", *Ann. N. Y. Acad. Sci.,* vol. 102, no. 1, pp. 29-45, 1962.
[http://dx.doi.org/10.1111/j.1749-6632.1962.tb13623.x] [PMID: 14021529]

[2]   A. Solaimuthu, A.N. Vijayan, P. Murali, and P.S. Korrapati, "Nano-biosensors and their relevance in tissue engineering", *Curr. Opin. Biomed. Eng.,* vol. 13, pp. 84-93, 2020.
[http://dx.doi.org/10.1016/j.cobme.2019.12.005]

[3]   S.K. Metkar, and K. Girigoswami, "Diagnostic biosensors in medicine – A review", *Biocatal. Agric. Biotechnol.,* vol. 17, pp. 271-283, 2019.
[http://dx.doi.org/10.1016/j.bcab.2018.11.029]

[4]   Q. Chen, J. Lin, C. Gan, Y. Wang, D. Wang, Y. Xiong, W. Lai, Y. Li, and M. Wang, "A sensitive impedance biosensor based on immunomagnetic separation and urease catalysis for rapid detection of *Listeria monocytogenes* using an immobilization-free interdigitated array microelectrode", *Biosens. Bioelectron.,* vol. 74, pp. 504-511, 2015.
[http://dx.doi.org/10.1016/j.bios.2015.06.007] [PMID: 26176211]

[5]   M. Rai, A. Gade, S. Gaikwad, P.D. Marcato, and N. Durán, "Biomedical applications of nanobiosensors: The state-of-the-art", *J. Braz. Chem. Soc.,* vol. 23, pp. 14-24, 2012.

[6]   R.C. Stringer, S. Schommer, D. Hoehn, and S.A. Grant, "Development of an optical biosensor using gold nanoparticles and quantum dots for the detection of Porcine Reproductive and Respiratory Syndrome Virus", *Sens. Actuators B Chem.,* vol. 134, no. 2, pp. 427-431, 2008.
[http://dx.doi.org/10.1016/j.snb.2008.05.018]

[7]   D. Dell'Atti, M. Zavaglia, S. Tombelli, G. Bertacca, A.O. Cavazzana, G. Bevilacqua, M. Minunni, and M. Mascini, "Development of combined DNA-based piezoelectric biosensors for the simultaneous detection and genotyping of high risk Human Papilloma Virus strains", *Clin. Chim. Acta,* vol. 383, no. 1-2, pp. 140-146, 2007.
[http://dx.doi.org/10.1016/j.cca.2007.05.009] [PMID: 17573061]

[8]   A. Mobed, B. Baradaran, M. Guardia, M. Agazadeh, M. Hasanzadeh, M.A. Rezaee, J. Mosafer, A. Mokhtarzadeh, and M.R. Hamblin, "Advances in detection of fastidious bacteria: From microscopic observation to molecular biosensors", *Trends Analyt. Chem.,* vol. 113, pp. 157-171, 2019.
[http://dx.doi.org/10.1016/j.trac.2019.02.012]

[9]   F. Malvano, R. Pilloton, and D. Albanese, "A novel impedimetric biosensor based on the antimicrobial activity of the peptide nisin for the detection of *Salmonella* spp", *Food Chem.,* vol. 325, p. 126868, 2020.
[http://dx.doi.org/10.1016/j.foodchem.2020.126868] [PMID: 32387945]

[10]   X. Hao, P. Yeh, Y. Qin, Y. Jiang, Z. Qiu, S. Li, T. Le, and X. Cao, "Aptamer surface functionalization of microfluidic devices using dendrimers as multi-handled templates and its application in sensitive detections of foodborne pathogenic bacteria", *Anal. Chim. Acta,* vol. 1056, pp. 96-107, 2019.
[http://dx.doi.org/10.1016/j.aca.2019.01.035] [PMID: 30797466]

[11]   A. Soni, R.K. Surana, and S.K. Jha, "Smartphone based optical biosensor for the detection of urea in saliva", *Sens. Actuators B Chem.,* vol. 269, pp. 346-353, 2018.
[http://dx.doi.org/10.1016/j.snb.2018.04.108]

[12]   A. Roda, E. Michelini, M. Zangheri, M. Di Fusco, D. Calabria, and P. Simoni, "Smartphone-based biosensors: A critical review and perspectives", *Trends Analyt. Chem.,* vol. 79, pp. 317-325, 2016.
[http://dx.doi.org/10.1016/j.trac.2015.10.019]

[13]   P. Mehrotra, "Biosensors and their applications – A review", *J. Oral Biol. Craniofac. Res.,* vol. 6, no. 2, pp. 153-159, 2016.
[http://dx.doi.org/10.1016/j.jobcr.2015.12.002] [PMID: 27195214]

[14]   W. Zhang, Y. Du, and M.L. Wang, "Noninvasive glucose monitoring using saliva nano-biosensor", *Sens. Biosensing Res.,* vol. 4, pp. 23-29, 2015.
[http://dx.doi.org/10.1016/j.sbsr.2015.02.002]

[15]   V.X.T. Zhao, T.I. Wong, X.T. Zheng, Y.N. Tan, and X. Zhou, "Colorimetric biosensors for point-of-care virus detections", *Mater. Sci. Energy Technol.,* vol. 3, pp. 237-249, 2020.
[http://dx.doi.org/10.1016/j.mset.2019.10.002] [PMID: 33604529]

[16]   A. Kordas, G. Papadakis, D. Milioni, J. Champ, S. Descroix, and E. Gizeli, "Rapid *Salmonella* detection using an acoustic wave device combined with the RCA isothermal DNA amplification method", *Sens. Biosensing Res.,* vol. 11, pp. 121-127, 2016.
[http://dx.doi.org/10.1016/j.sbsr.2016.10.010]

[17]   A. Hanif, R. Farooq, M.U. Rehman, R. Khan, S. Majid, and M.A. Ganaie, "Aptamer based nanobiosensors: Promising healthcare devices", *Saudi Pharm. J.,* vol. 27, no. 3, pp. 312-319, 2019.
[http://dx.doi.org/10.1016/j.jsps.2018.11.013] [PMID: 30976173]

[18]   P.R. Solanki, M.K. Patel, A. Kaushik, M.K. Pandey, R.K. Kotnala, and B.D. Malhotra, "Sol-Gel Derived Nanostructured Metal Oxide Platform for Bacterial Detection", *Electroanalysis,* vol. 23, no. 11, pp. 2699-2708, 2011.
[http://dx.doi.org/10.1002/elan.201100351]

[19]   S. Samanman, P. Kanatharana, W. Chotigeat, P. Deachamag, and P. Thavarungkul, "Highly sensitive capacitive biosensor for detecting white spot syndrome virus in shrimp pond water", *J. Virol. Methods,* vol. 173, no. 1, pp. 75-84, 2011.
[http://dx.doi.org/10.1016/j.jviromet.2011.01.010] [PMID: 21256870]

[20]   N. Bhalla, P. Jolly, N. Formisano, and P. Estrela, "Introduction to biosensors", *Essays Biochem.,* vol. 60, no. 1, pp. 1-8, 2016.
[http://dx.doi.org/10.1042/EBC20150001] [PMID: 27365030]

[21]   N. Khansili, G. Rattu, and P.M. Krishna, "Label-free optical biosensors for food and biological sensor applications", *Sens. Actuators B Chem.,* vol. 265, pp. 35-49, 2018.
[http://dx.doi.org/10.1016/j.snb.2018.03.004]

[22]   V.N. Konopsky, and E.V. Alieva, "Imaging biosensor based on planar optical waveguide", *Opt. Laser Technol.,* vol. 115, pp. 171-175, 2019.
[http://dx.doi.org/10.1016/j.optlastec.2019.02.034]

[23]   P. Vidal, M. Martinez, C. Hernandez, A.R. Adhikari, Y. Mao, L. Materon, and K. Lozano, "Development of chromatic biosensor for quick bacterial detection based on polyvinyl butyrate-polydiacetylene nonwoven fiber composites", *Eur. Polym. J.,* vol. 121, p. 109284, 2019.
[http://dx.doi.org/10.1016/j.eurpolymj.2019.109284]

[24]   W. Jeong, S.H. Choi, H. Lee, and Y. Lim, "A fluorescent supramolecular biosensor for bacterial detection *via* binding-induced changes in coiled-coil molecular assembly", *Sens. Actuators B Chem.,* vol. 290, pp. 93-99, 2019.
[http://dx.doi.org/10.1016/j.snb.2019.03.112]

[25]   M. Mathelié-Guinlet, T. Cohen-Bouhacina, I. Gammoudi, A. Martin, L. Béven, M.H. Delville, and C. Grauby-Heywang, "Silica nanoparticles-assisted electrochemical biosensor for the rapid, sensitive and specific detection of *Escherichia coli*", *Sens. Actuators B Chem.,* vol. 292, pp. 314-320, 2019.
[http://dx.doi.org/10.1016/j.snb.2019.03.144]

[26]   S.H. Baek, M.W. Kim, C.Y. Park, C.S. Choi, S.K. Kailasa, J.P. Park, and T.J. Park, "Development of a rapid and sensitive electrochemical biosensor for detection of human norovirus *via* novel specific binding peptides", *Biosens. Bioelectron.,* vol. 123, pp. 223-229, 2019.
[http://dx.doi.org/10.1016/j.bios.2018.08.064] [PMID: 30195404]

[27]   M. Pohanka, "Overview of piezoelectric biosensors, immunosensors and DNA sensors and their applications", *Materials (Basel),* vol. 11, no. 3, p. 448, 2018.
[http://dx.doi.org/10.3390/ma11030448] [PMID: 29562700]

[28]   M.J. Linman, K. Sugerman, and Q. Cheng, "Detection of low levels of *Escherichia coli* in fresh spinach by surface plasmon resonance spectroscopy with a TMB-based enzymatic signal enhancement method. Sensors Actuators", *Biol. Chem.,* vol. 145, pp. 613-619, 2010.
[http://dx.doi.org/10.1016/j.snb.2010.01.007]

[29]   X. Wei, W. Zhou, S.T. Sanjay, J. Zhang, Q. Jin, F. Xu, D.C. Dominguez, and X. Li, "F. "XuMultiplexed instrument-free bar-chart SpinChip integrated with nanoparticle-mediated magnetic aptasensors for visual quantitative detection of multiple pathogens,"", *Anal. Chem.,* vol. 90, no. 16, pp. 9888-9896, 2018.
[http://dx.doi.org/10.1021/acs.analchem.8b02055] [PMID: 30028601]

[30]   J.S. Kee, S.Y. Lim, A.P. Perera, Y. Zhang, and M.K. Park, "Plasmonic nanohole arrays for monitoring growth of bacteria and antibiotic susceptibility test", *Sens. Actuators B Chem.,* vol. 182, pp. 576-583, 2013.
[http://dx.doi.org/10.1016/j.snb.2013.03.053]

[31]   M.S. Khan, S.K. Misra, K. Dighe, Z. Wang, A.S. Schwartz-Duval, D. Sar, and D. Pan, "Electrically-receptive and thermally-responsive paper-based sensor chip for rapid detection of bacterial cells", *Biosens. Bioelectron.,* vol. 110, pp. 132-140, 2018.
[http://dx.doi.org/10.1016/j.bios.2018.03.044] [PMID: 29605712]

[32]   P. Poltronieri, F. Cimaglia, E. De Lorenzis, M. Chiesa, V. Mezzolla, and I. Reca, "Protein chips for detection of *Salmonella* spp. from enrichment culture", *Sensors (Basel),* vol. 16, no. 4, p. 574, 2016.
[http://dx.doi.org/10.3390/s16040574] [PMID: 27110786]

[33]   H. Tachibana, M. Saito, S. Shibuya, K. Tsuji, N. Miyagawa, K. Yamanaka, and E. Tamiya, "On-chip quantitative detection of pathogen genes by autonomous microfluidic PCR platform", *Biosens. Bioelectron.,* vol. 74, pp. 725-730, 2015.
[http://dx.doi.org/10.1016/j.bios.2015.07.009] [PMID: 26210470]

[34]   S.J. Oh, B.H. Park, J.H. Jung, G. Choi, D.C. Lee, D.H. Kim, and T.S. Seo, "Centrifugal loop-mediated isothermal amplification microdevice for rapid, multiplex and colorimetric foodborne pathogen detection", *Biosens. Bioelectron.,* vol. 75, pp. 293-300, 2016.

[http://dx.doi.org/10.1016/j.bios.2015.08.052] [PMID: 26322592]

[35] S. Bouguelia, Y. Roupioz, S. Slimani, L. Mondani, M.G. Casabona, C. Durmort, T. Vernet, R. Calemczuk, and T. Livache, "On-chip microbial culture for the specific detection of very low levels of bacteria", *Lab Chip,* vol. 13, no. 20, pp. 4024-4032, 2013.
[http://dx.doi.org/10.1039/c3lc50473e] [PMID: 23912527]

[36] D.D. Galvan, V. Parekh, E. Liu, E.L. Liu, and Q. Yu, "Sensitive bacterial detection *via* dielectrophoretic-enhanced mass transport using surface-plasmon-resonance biosensors", *Anal. Chem.,* vol. 90, no. 24, pp. 14635-14642, 2018.
[http://dx.doi.org/10.1021/acs.analchem.8b05137] [PMID: 30433764]

[37] C. Xue, C. Yang, T. Xu, J. Zhan, and X. Li, "A wireless bio-sensing microfluidic chip based on resonating 'μ-divers'", *Lab Chip,* vol. 15, no. 10, pp. 2318-2326, 2015.
[http://dx.doi.org/10.1039/C5LC00361J] [PMID: 25891094]

[38] B. Koo, A.M. Yorita, J.J. Schmidt, and H.G. Monbouquette, "Amplification-free, sequence-specific 16S rRNA detection at 1 aM", *Lab Chip,* vol. 18, no. 15, pp. 2291-2299, 2018.
[http://dx.doi.org/10.1039/C8LC00452H] [PMID: 29987290]

[39] A. Pandey, Y. Gurbuz, V. Ozguz, J.H. Niazi, and A. Qureshi, "Graphene-interfaced electrical biosensor for label-free and sensitive detection of foodborne pathogenic *E. coli* O157:H7", *Biosens. Bioelectron.,* vol. 91, pp. 225-231, 2017.
[http://dx.doi.org/10.1016/j.bios.2016.12.041] [PMID: 28012318]

[40] A. Bajwa, S. Tan, R. Mehta, and B. Bahreyni, "Rapid detection of viable microorganisms based on a plate count technique using arrayed microelectrodes", *Sensors (Basel),* vol. 13, no. 7, pp. 8188-8198, 2013.
[http://dx.doi.org/10.3390/s130708188] [PMID: 23803788]

[41] S.J. Oh, and T.S. Seo, "Combination of a centrifugal microfluidic device with a solution-loading cartridge for fully automatic molecular diagnostics", *Analyst (Lond.),* vol. 144, no. 19, pp. 5766-5774, 2019.
[http://dx.doi.org/10.1039/C9AN00900K] [PMID: 31436781]

[42] J.R. Choi, J. Hu, R. Tang, Y. Gong, S. Feng, H. Ren, T. Wen, X. Li, W.A.B. Wan Abas, B. Pingguan-Murphy, and F. Xu, "An integrated paper-based sample-to-answer biosensor for nucleic acid testing at the point of care", *Lab Chip,* vol. 16, no. 3, pp. 611-621, 2016.
[http://dx.doi.org/10.1039/C5LC01388G] [PMID: 26759062]

[43] C. RoyChaudhuri, and R. Dev Das, "A biomolecule compatible electrical model of microimpedance affinity biosensor for sensitivity improvement in cell detection", *Sens. Actuators A Phys.,* vol. 157, no. 2, pp. 280-289, 2010.
[http://dx.doi.org/10.1016/j.sna.2009.11.031]

[44] B. Heidenreich, C. Poehlmann, M. Sprinzl, and M. Gareis, "Rapid detection of bacteria by biochips during food processing", *Fleischwirtschaft (Frankf.),* vol. 90, no. 7, pp. 96-99, 2013.

[45] A. Sayad, F. Ibrahim, S. Mukim Uddin, J. Cho, M. Madou, and K.L. Thong, "A microdevice for rapid, monoplex and colorimetric detection of foodborne pathogens using a centrifugal microfluidic platform", *Biosens. Bioelectron.,* vol. 100, pp. 96-104, 2018.
[http://dx.doi.org/10.1016/j.bios.2017.08.060] [PMID: 28869845]

[46] G. Kim, J.H. Moon, C.Y. Moh, and J. Lim, "A microfluidic nano-biosensor for the detection of pathogenic *Salmonella*", *Biosens. Bioelectron.,* vol. 67, pp. 243-247, 2015.
[http://dx.doi.org/10.1016/j.bios.2014.08.023] [PMID: 25172028]

[47] L. Wang, R. Wang, F. Chen, T. Jiang, H. Wang, M. Slavik, H. Wei, and Y. Li, "QCM-based aptamer selection and detection of *Salmonella typhimurium*", *Food Chem.,* vol. 221, pp. 776-782, 2017.
[http://dx.doi.org/10.1016/j.foodchem.2016.11.104] [PMID: 27979272]

[48] T.Q. Hung, W.H. Chin, Y. Sun, A. Wolff, and D.D. Bang, "A novel lab-on-chip platform with

integrated solid phase PCR and Supercritical Angle Fluorescence (SAF) microlens array for highly sensitive and multiplexed pathogen detection", *Biosens. Bioelectron.,* vol. 90, pp. 217-223, 2017.
[http://dx.doi.org/10.1016/j.bios.2016.11.028] [PMID: 27902940]

[49] I. Jasim, Z. Shen, Z. Mlaji, N.S. Yuksek, A. Abdullah, J. Liu, S.G. Dastider, M. El-Dweik, S. Zhang, and M. Almasri, "An impedance biosensor for simultaneous detection of low concentration of *Salmonella serogroups* in poultry and fresh produce samples", *Biosens. Bioelectron.,* vol. 126, pp. 292-300, 2019.
[http://dx.doi.org/10.1016/j.bios.2018.10.065] [PMID: 30445304]

[50] P.D. Nguyen, T.B. Tran, D.T.X. Nguyen, and J. Min, "Magnetic silica nanotube-assisted impedimetric immunosensor for the separation and label-free detection of *Salmonella typhimurium*", *Sens. Actuators B Chem.,* vol. 197, pp. 314-320, 2014.
[http://dx.doi.org/10.1016/j.snb.2014.02.089]

[51] H. Zhang, L. Xue, F. Huang, S. Wang, L. Wang, N. Liu, and J. Lin, "A capillary biosensor for rapid detection of *Salmonella* using Fe-nanocluster amplification and smart phone imaging", *Biosens. Bioelectron.,* vol. 127, pp. 142-149, 2019.
[http://dx.doi.org/10.1016/j.bios.2018.11.042] [PMID: 30597432]

[52] B. Park, and S.J. Choi, "Sensitive immunoassay-based detection of *Vibrio parahaemolyticus* using capture and labeling particles in a stationary liquid phase lab-on-a-chip", *Biosens. Bioelectron.,* vol. 90, pp. 269-275, 2017.
[http://dx.doi.org/10.1016/j.bios.2016.11.071] [PMID: 27923189]

[53] Y.M. Park, S.Y. Lim, S.J. Shin, C.H. Kim, S.W. Jeong, S.Y. Shin, N.H. Bae, S.J. Lee, J. Na, G.Y. Jung, and T.J. Lee, "A film-based integrated chip for gene amplification and electrochemical detection of pathogens causing foodborne illnesses", *Anal. Chim. Acta,* vol. 1027, pp. 57-66, 2018.
[http://dx.doi.org/10.1016/j.aca.2018.03.061] [PMID: 29866270]

[54] T.T. Nguyen, K.T.L. Trinh, W.J. Yoon, N.Y. Lee, and H. Ju, "Integration of a microfluidic polymerase chain reaction device and surface plasmon resonance fiber sensor into an inline all-in-one platform for pathogenic bacteria detection", *Sens. Actuators B Chem.,* vol. 242, pp. 1-8, 2017.
[http://dx.doi.org/10.1016/j.snb.2016.10.137]

[55] A.C. Vinayaka, T.A. Ngo, K. Kant, P. Engelsmann, V.P. Dave, M.A. Shahbazi, A. Wolff, and D.D. Bang, "Rapid detection of *Salmonella enterica* in food samples by a novel approach with combination of sample concentration and direct PCR", *Biosens. Bioelectron.,* vol. 129, pp. 224-230, 2019.
[http://dx.doi.org/10.1016/j.bios.2018.09.078] [PMID: 30318404]

[56] S. Wang, L. Zheng, G. Cai, N. Liu, M. Liao, Y. Li, X. Zhang, and J. Lin, "A microfluidic biosensor for online and sensitive detection of *Salmonella typhimurium* using fluorescence labeling and smartphone video processing", *Biosens. Bioelectron.,* vol. 140, p. 111333, 2019.
[http://dx.doi.org/10.1016/j.bios.2019.111333] [PMID: 31153017]

[57] S. Alhogail, G.A.R.Y. Suaifan, and M. Zourob, "Rapid colorimetric sensing platform for the detection of *Listeria monocytogenes* foodborne pathogen", *Biosens. Bioelectron.,* vol. 86, pp. 1061-1066, 2016.
[http://dx.doi.org/10.1016/j.bios.2016.07.043] [PMID: 27543841]

[58] R. Sidhu, Y. Rong, D.C. Vanegas, J. Claussen, E.S. McLamore, and C. Gomes, "Impedance biosensor for the rapid detection of Listeria spp. based on aptamer functionalized Pt-interdigitated microelectrodes array", *Proceedings of the SPIE,* vol. 9863, p. 8, 2016.
[http://dx.doi.org/10.1117/12.2223443]

[59] Q. Chen, D. Wang, G. Cai, Y. Xiong, Y. Li, M. Wang, H. Huo, and J. Lin, "Fast and sensitive detection of foodborne pathogen using electrochemical impedance analysis, urease catalysis and microfluidics", *Biosens. Bioelectron.,* vol. 86, pp. 770-776, 2016.
[http://dx.doi.org/10.1016/j.bios.2016.07.071] [PMID: 27476059]

[60] P.P. Sharma, E. Albisetti, M. Massetti, M. Scolari, C. La Torre, M. Monticelli, M. Leone, F. Damin, G. Gervasoni, G. Ferrari, F. Salice, E. Cerquaglia, G. Falduti, M. Cretich, E. Marchisio, M. Chiari, M.

Sampietro, D. Petti, and R. Bertacco, "Integrated platform for detecting pathogenic DNA *via* magnetic tunneling junction-based biosensors", *Sens. Actuators B Chem.*, vol. 242, pp. 280-287, 2017.
[http://dx.doi.org/10.1016/j.snb.2016.11.051]

[61]   K. Pisamayarom, A. Suriyasomboon, and P. Chaumpluk, "Simple screening of *Listeria monocytogenes* based on a fluorescence assay *via* a laminated lab-on-paper chip", *Biosensors (Basel)*, vol. 7, no. 4, p. 56, 2017.
[http://dx.doi.org/10.3390/bios7040056] [PMID: 29182562]

[62]   M. Ali, A.R. Nelson, A.L. Lopez, and D.A. Sack, "Updated global burden of cholera in endemic countries", *PLoS Negl. Trop. Dis.*, vol. 9, no. 6, p. e0003832, 2015.
[http://dx.doi.org/10.1371/journal.pntd.0003832] [PMID: 26043000]

[63]   Y. Peng, Y. Jin, H. Lin, J. Wang, and M.N. Khan, "Application of the VPp1 bacteriophage combined with a coupled enzyme system in the rapid detection of *Vibrio parahaemolyticus*", *J. Microbiol. Methods*, vol. 98, pp. 99-104, 2014.
[http://dx.doi.org/10.1016/j.mimet.2014.01.005] [PMID: 24440165]

[64]   C. Duarte, T. Costa, C. Carneiro, R. Soares, A. Jitariu, S. Cardoso, M. Piedade, R. Bexiga, and P. Freitas, "Semi-quantitative method for *streptococci* magnetic detection in raw milk", *Biosensors (Basel)*, vol. 6, no. 2, p. 19, 2016.
[http://dx.doi.org/10.3390/bios6020019] [PMID: 27128950]

[65]   P.B. Lillehoj, C.W. Kaplan, J. He, W. Shi, and C.M. Ho, "Rapid, electrical impedance detection of bacterial pathogens using immobilized antimicrobial peptides", *SLAS Technol.*, vol. 19, no. 1, pp. 42-49, 2014.
[http://dx.doi.org/10.1177/2211068213495207] [PMID: 23850865]

[66]   A.C. Fernandes, C.M. Duarte, F.A. Cardoso, R. Bexiga, S. Cardoso, and P.P. Freitas, "Lab-on-chip cytometry based on magnetoresistive sensors for bacteria detection in milk", *Sensors (Basel).*, vol. 14, no. 8, pp. 15496-524, 2014.
[http://dx.doi.org/10.3390/s140815496] [PMID: 25196163]

[67]   S. Poshtiban, A. Singh, G. Fitzpatrick, and S. Evoy, "Bacteriophage tail-spike protein derivitized microresonator arrays for specific detection of pathogenic bacteria", *Sens. Actuators B Chem.*, vol. 181, pp. 410-416, 2013.
[http://dx.doi.org/10.1016/j.snb.2012.12.082]

[68]   N.M.M. Pires, and T. Dong, "An integrated passive-flowmicrofluidic biosensor with organic photodiodes for ultra-sensitive pathogen detection in water", *Proc. of the 2014 36th Ann. Inter. Conf. of the IEEE Engi. in Medi. and Biol. Soci.*, pp. 4411-4414, 2014.

[69]   Z. Ölcer, E. Esen, A. Ersoy, S. Budak, D. Sever Kaya, M. Yağmur Gök, S. Barut, D. Üstek, and Y. Uludag, "Microfluidics and nanoparticles based amperometric biosensor for the detection of cyanobacteria (Planktothrix agardhii NIVA-CYA 116) DNA", *Biosens. Bioelectron.*, vol. 70, pp. 426-432, 2015.
[http://dx.doi.org/10.1016/j.bios.2015.03.052] [PMID: 25845335]

[70]   B.A. Adducci, H.A. Gruszewski, P.A. Khatibi, and D.G. Schmale III, "Differential detection of a surrogate biological threat agent (*Bacillus globigii*) with a portable surface plasmon resonance biosensor", *Biosens. Bioelectron.*, vol. 78, pp. 160-166, 2016.
[http://dx.doi.org/10.1016/j.bios.2015.11.032] [PMID: 26606307]

[71]   C.M. Duarte, C. Carneiro, S. Cardoso, P.P. Freitas, and R. Bexiga, "Semi-quantitative method for *Staphylococci* magnetic detection in raw milk", *J. Dairy Res.*, vol. 84, no. 1, pp. 80-88, 2017.
[http://dx.doi.org/10.1017/S0022029916000741] [PMID: 28007038]

[72]   L.M. De Plano, S. Carnazza, G.M.L. Messina, M.G. Rizzo, G. Marletta, and S.P.P. Guglielmino, "Specific and selective probes for *Staphylococcus aureus* from phage-displayed random peptide libraries", *Colloids Surf. B Biointerfaces*, vol. 157, pp. 473-480, 2017.
[http://dx.doi.org/10.1016/j.colsurfb.2017.05.081] [PMID: 28654884]

[73]   K.T.L. Trinh, T.N.D. Trinh, and N.Y. Lee, "Fully integrated and slidable paper-embedded plastic microdevice for point-of-care testing of multiple foodborne pathogens", *Biosens. Bioelectron.,* vol. 135, pp. 120-128, 2019.
[http://dx.doi.org/10.1016/j.bios.2019.04.011] [PMID: 31004922]

[74]   D. Tang, J. Tang, B. Su, J. Ren, and G. Chen, "Simultaneous determination of five-type hepatitis virus antigens in 5min using an integrated automatic electrochemical immunosensor array", *Biosens. Bioelectron.,* vol. 25, no. 7, pp. 1658-1662, 2010.
[http://dx.doi.org/10.1016/j.bios.2009.12.004] [PMID: 20022741]

[75]   S.A. Hong, J. Kwon, D. Kim, and S. Yang, "A rapid, sensitive and selective electrochemical biosensor with concanavalin A for the preemptive detection of norovirus", *Biosens. Bioelectron.,* vol. 64, pp. 338-344, 2015.
[http://dx.doi.org/10.1016/j.bios.2014.09.025] [PMID: 25254625]

[76]   M. Kitajima, N. Wang, M.Q.X. Tay, J. Miao, and A.J. Whittle, "Development of a MEMS-based electrochemical aptasensor for norovirus detection", *Micro – Nano Lett.,* vol. 11, no. 10, pp. 582-585, 2016.
[http://dx.doi.org/10.1049/mnl.2016.0295]

[77]   A. Panghal, and S.J. Flora, Chapter 4—Viral agents including threat from emerging viral infections. *Handbook on Biological Warfare Preparedness.,* S.J. Flora, V. Pachauri, Eds., Academic Press: Cambridge, MA, USA. 65-81, 2020.
[http://dx.doi.org/10.1016/B978-0-12-812026-2.00004-9]

[78]   M.Z.H. Khan, M.R. Hasan, S.I. Hossain, M.S. Ahommed, and M. Daizy, "Ultrasensitive detection of pathogenic viruses with electrochemical biosensor: State of the art", *Biosens. Bioelectron.,* vol. 166, p. 112431, 2020.
[http://dx.doi.org/10.1016/j.bios.2020.112431] [PMID: 32862842]

[79]   X. Yuan, C. Yang, Q. He, J. Chen, D. Yu, J. Li, S. Zhai, Z. Qin, K. Du, Z. Chu, and P. Qin, "Current and perspective diagnostic techniques for COVID-19", *ACS Infect. Dis.,* vol. 6, no. 8, pp. 1998-2016, 2020.
[http://dx.doi.org/10.1021/acsinfecdis.0c00365] [PMID: 32677821]

[80]   M. Kaur, S. Tiwari, and R. Jain, "Protein-based biomarkers for non-invasive Covid-19 detection. Sens," Bio-Sens", *Res.,* vol. 29, pp. 100-162, 2020.

[81]   J. Zhifeng, A. Feng, and T. Li, "Consistency analysis of COVID-19 nucleic acid tests and the changes of lung CT", *J. Clin. Virol.,* vol. 127, p. 104359, 2020.
[http://dx.doi.org/10.1016/j.jcv.2020.104359] [PMID: 32302956]

[82]   "FDA. Health C for D and R. EUA Authorized Serology Test Performance", Available at: https://www.fda.gov/medical-devices/coronavirus-disease-2019-covid-19-emerge-cy-use-authorizationsmedical-devices/eua-authorized-serology-test-performance (accessed on: 2020).

[83]   E. Morales, and C. Dincer, "The impact of bio-sensing in a pandemic outbreak: COVID-19,'", In: *Biosens. Bioelectron.* vol. 163. Elsevier. 112-274, 2020.
[http://dx.doi.org/10.1016/j.bios.2020.112274]

[84]   J. Yin, Z. Zou, Z. Hu, S. Zhang, F. Zhang, B. Wang, S. Lv, and Y. Mu, "A "sample-in-multipl-x-digital-answer-out" chip for fast detection of pathogens", *Lab Chip,* vol. 20, no. 5, pp. 979-986, 2020.
[http://dx.doi.org/10.1039/C9LC01143A] [PMID: 32003380]

[85]   N.C. Cady, V. Fusco, G. Maruccio, E. Primiceri, and C.A. Batt, "Micro- and nanotechnology-based approaches to detect pathogenic agents in food", In: *Nanobiosensors.,* A.M. Grumezescu, Ed., Academic Press: Cambridge, MA, USA, 2017, pp. 475-510.
[http://dx.doi.org/10.1016/B978-0-12-804301-1.00012-6]

[86]   J. Zhuang, J. Yin, S. Lv, B. Wang, and Y. Mu, "Advanced "lab-on-a-chip" to detect viruses – Current

challenges and future perspectives", *Biosens. Bioelectron.,* vol. 163, p. 112291, 2020.
[http://dx.doi.org/10.1016/j.bios.2020.112291] [PMID: 32421630]

[87]   S. Mavrikou, G. Moschopoulou, V. Tsekouras, and S. Kintzios, "Development of a portable, ultra-rapid and ultra-sensitive cell-based biosensor for the direct detection of the SARS-CoV-2 S1 spike protein antigen", *Sensors (Basel),* vol. 20, no. 11, p. 3121, 2020.
[http://dx.doi.org/10.3390/s20113121] [PMID: 32486477]

[88]   C. Choi, "Integrated nanobiosensor technology for biomedical application", *Nanobiosens. Dis. Diagn.,* vol. 1, pp. 1-4, 2012.
[http://dx.doi.org/10.2147/NDD.S26422]

[89]   G. Seo, G. Lee, M.J. Kim, S.H. Baek, M. Choi, K.B. Ku, C.S. Lee, S. Jun, D. Park, H.G. Kim, S.J. Kim, J.O. Lee, B.T. Kim, E.C. Park, and S.I. Kim, "Rapid detection of COVID-19 causative virus (SARS-CoV-2) in human nasopharyngeal swab specimens using field-effect transistor-based biosensor", *ACS Nano,* vol. 14, no. 4, pp. 5135-5142, 2020.
[http://dx.doi.org/10.1021/acsnano.0c02823] [PMID: 32293168]

[90]   N. Bhalla, Y. Pan, Z. Yang, and A.F. Payam, "Opportunities and challenges for biosensors and nanoscale analytical tools for pandemics: COVID-19", *ACS Nano,* vol. 14, no. 7, pp. 7783-7807, 2020.
[http://dx.doi.org/10.1021/acsnano.0c04421] [PMID: 32551559]

[91]   C. Ricciardi, G. Canavese, R. Castagna, G. Digregorio, I. Ferrante, S.L. Marasso, A. Ricci, V. Alessandria, K. Rantsiou, and L.S. Cocolin, "Online portable microcantilever biosensors for *Salmonella enterica serotype enteritidis* detection", *Food Bioprocess Technol.,* vol. 3, no. 6, pp. 956-960, 2010.
[http://dx.doi.org/10.1007/s11947-010-0362-0]

[92]   M.C. Morant-Miñana, and J. Elizalde, "Microscale electrodes integrated on COP for real sample Campylobacter spp. detection", *Biosens. Bioelectron.,* vol. 70, pp. 491-497, 2015.
[http://dx.doi.org/10.1016/j.bios.2015.03.063] [PMID: 25889259]

[93]   L. Xu, R. Wang, L.C. Kelso, Y. Ying, and Y. Li, "A target-responsive and size-dependent hydrogel aptasensor embedded with QD fluorescent reporters for rapid detection of avian influenza virus H5N1", *Sens. Actuators B Chem.,* vol. 234, pp. 98-108, 2016.
[http://dx.doi.org/10.1016/j.snb.2016.04.156]

[94]   F. Zang, K. Gerasopoulos, X.Z. Fan, A.D. Brown, J.N. Culver, and R. Ghodssi, "Real-time monitoring of macromolecular biosensing probe self-assembly and on-chip ELISA using impedimetric microsensors", *Biosens. Bioelectron.,* vol. 81, pp. 401-407, 2016.
[http://dx.doi.org/10.1016/j.bios.2016.03.019] [PMID: 26995286]

CHAPTER 6

Biosensor: A Technology that Changed the way of Diabetes Diagnosis

Sushil Kumar Dubey[1,*], **Bhaskar Sharma**[2], **Divya Mishra**[3] and **M.P. Singh**[1]

[1] *Centre of Biotechnology, University of Allahabad., Prayagraj, Uttar Pradesh, 211002, India*

[2] *Neurobiology Laboratory, Department of Anatomy, All India Institute of Medical Sciences, New Delhi, 110049, India*

[3] *Centre of Bioinformatics, University of Allahabad, Prayagraj, Uttar Pradesh, 211002, India*

Abstract: Over the past few decades, technology has greatly improved, with more accuracy and efficiency in monitoring the blood glucose level of diabetic patients. To monitor the blood glucose level, several glucose biosensors have been developed. However, the accuracy, efficiency, standardization and training of the users in an easy way is still a challenge. Biosensors are a non-invasive method that helps medical workers to find the required doses of insulin for diabetic patients. In this chapter, we will discuss different types of biosensors developed, their working mechanism, and the current scenario of biosensors in diabetes diagnosis.

Keywords: Biosensors, Diabetes, Glucose Biosensors, Mechanism.

INTRODUCTION

Diabetes is a serious and complex health problem in the 21st century. Globally, there will be 422 million people with diabetes as of 2014, many of whom live in low- and middle-income countries. The number of people with diabetes has been steadily increasing over the decade. It also has been estimated to affect 552 million people around the globe by 2030 [1]. Diabetes is a major cause of other severe diseases, including heart and kidney disease, and can also cause blindness in the elderly, yet almost half of the diabetes-related cases remain undiagnosed. Diabetes is one of the leading causes of death worldwide, having 1.6 million death per year, and its research has never been more significant [2]. To reduce the number of deaths caused by diabetes, the WHO has listed the priority medical devices to manage cardiovascular diseases and diabetes (**released on 30 June 2021**), which will help healthcare providers and policymakers prioritize the procurement of medical devices for diabetic patients. The list consists of hundreds

---

[*] **Corresponding author Sushil Kumar Dubey:** Centre of Biotechnology, University of Allahabad, Prayagraj, Uttar Pradesh–211002 India; Email: sushildubey40@gmail.com

Vivek K. Chaturvedi, Dawesh P. Yadav and Mohan P. Singh (Eds.)

of medical devices that are mainly required for the detection and treatment of diabetes. It starts from primary care facilities to extremely specialized hospitals and medical devices that are needed for health-related emergencies (such as hypo or hyperglycemic emergencies, WHO 2021).

According to the reports suggested by Diabetes Australia, around 12% of global health costs are spent on diabetes, which is USD 673 billion [2]. Therefore, it is significant to focus on effective diagnostic tools and low-cost monitoring devices for diabetes. There are mainly three types of diabetes: type 1, type 2 and gestational diabetes. Therefore, type 1 diabetes mellitus is a chronic metabolic disorder that is characterized by high fasting glucose, in which beta cells (from islets of Langerhans in the pancreas) can no longer provide insulin. Insulin is needed to maintain the level of blood glucose in the range of 4.0-5.5 mmol/L [3]. Likewise, it is suggested that when the body fails to maintain the optimum range of blood glucose, it leads to hypo or hyperglycaemia. The condition of hypo or hyperglycaemia will further lead to physiological complications and provide severe damage to major organs. In the case of an undiagnosed diabetic patient, the type 1 diabetes symptoms include polydipsia (thirst), polyuria (excessive urination), weight loss, constant hunger, and fatigue, among others [4]. According to research, type 1 diabetes cannot be preventable at this time, but its monitoring is of utmost significance for diabetic patients. Type 2 diabetes is the result of the degeneration or lack of insulin response. However, type 2 diabetes is the most common type of diabetes in the world today, representing 85 to 90% of all cases [1]. Type 1 diabetes is more readily managed with regular diet, exercise, and the control of lipids and blood pressure [4], although insulin can still be needed. However, further prevention from developing type 2 diabetes will significantly decrease the risk of complications, *i.e.*, blindness, kidney failure, and cardiovascular disease. Evidence suggests that type 2 diabetes can be cured through accurate management of dysregulated insulin and glucose [5]. Hence, real-time glucose measurement is significant for preventing and predicting glucose levels and following medications to manage diabetes. Certainly, gestational diabetes mainly occurs during the period of pregnancy, affecting between 12 and 14% of pregnant women. Most gestational pregnancies are healthy pregnancies and produce healthy offspring if they follow good eating habits, and most women will no longer have high blood sugar after the baby is born [1]. Thus, close glucose level measurements are highly recommended by healthcare specialists to guide proper exercise and healthy food for pregnant women. Currently, the most popular and significant approach for diabetes management is focused on at-home glucose monitoring using the finger-pricking method. In the finger pricking method, the patients must maintain their daytime sugar levels between 4.4 and 6.7 mm/L, using a glucose meter. Although this measurement approach can become a burden, that relies on a high frequency of glucose monitoring to result in good glycaemic control [6], and it doesn't warn of the occurrences of hypoglycaemia. While testing, the wounds will cause through needle sticks which contain the risk of infection. In the case of children patients, the pain may cause fear, avoid blood glucose monitoring, and they can refuse medical treatment. Also, this approach can lead to loss of fingertips sensations with the occurrence of pain [3]. The finger-pricking testing is also known to serve unreliable data because of inaccurate blood sampling and lead to infections due to regular intrusive testing.

An alternative approach is the non-intrusive continuous glucose testing approach that provides a comprehensive insight into the patient's glucose monitoring and refuses medical treatment. However, research has been done to compare continuous glucose monitoring with or without regular blood glucose measurements and confirmed that it is as efficient and safe to use in adults with type 1 diabetes with a low risk of hypoglycaemia [7].

Evidence suggests that by using continuous glucose monitoring (CGM), patients can essentially reduce their HbA1c levels as opposed to routine care. In addition, CGM medical devices provide better glucose control and reduce the frequency of insulin management; also, more analyses are required to confirm this. With the emergence of chemistry, materials science, computer science, and wearable and wireless technology has made an essential advancement in digital quantifications of our daily routine, respectively. It needed a flexible substrate, an efficient sensor, and a reliable signal report system [8]. The current pandemic situations aid the market demand for disease testing, prevention, and management devices. It also leads the production of wearable sensing devices, which further alleviate an emerging field of research and matching proudly forward from measuring physical activity to health conditions [9]. Wearables have a wide range of uses in healthcare due to their point-of-care nature and non-invasiveness. Wearable medical devices can offer a platform for regular monitoring and recording of physiological information in the form of a device, which can be utilized for tracking patients' current medical situations, diagnoses, and treatment. However, the most common wearable sensor device is a smartwatch that can track regular activities and physiological parameters like temperature, heart rate, as well as sleep cycles. Likewise, wearable devices can also be utilized in the telehealth context that can efficiently measure patients, predict diseases as well as remind wearers to attend appointments by measuring the chemical information, that is, glucose and metal ions, and oxygen levels in our body fluids. This chapter summarises recent ongoing research activities based on the different biosensors for regular glucose measurements over the last 5 years.

Glucose Monitoring by Using the Biosensors

The biosensor is defined as a "compact analytical device or unit incorporating a biological or biologically derived sensitive recognition element integrated or associated with a physio-chemical transducer" [10]. This biosensor can be categorized mainly in three basic parts, *i.e.,*

a. The element which recognizes the target biological molecules among various chemical compounds,
b. The transducer element that converts the recognized signals into measurable form, and
c. A system that converts all signals into readable data [11, 12, 13].

The biomolecule recognition system targets receptors, microorganisms, enzymes, nucleic acids, lectins, and antibodies (Fig. **1**) [14, 15]. There are five main transducer classes which are magnetic, electrochemical, thermometric, and piezoelectric [16].

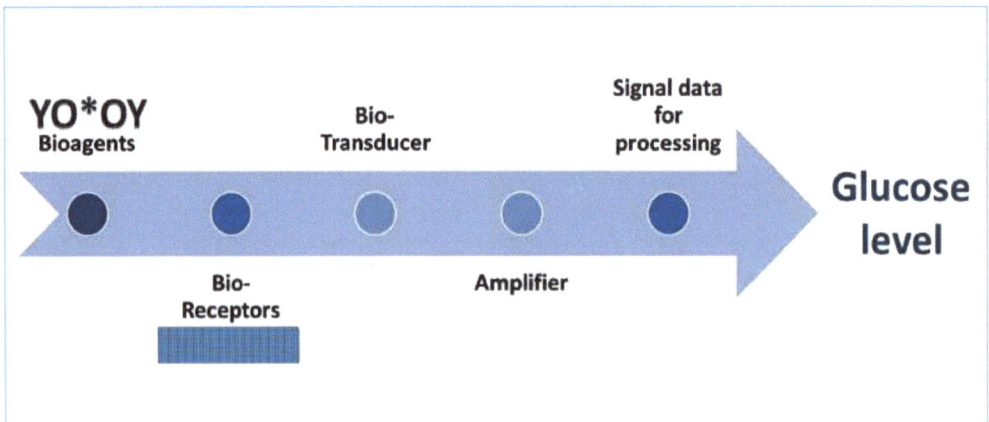

Fig. (1). The basic principle of biosensors.

The most efficient biosensors are of the electrochemical class because of their low cost, ease to handle, maintenance, reproducibility, and sensitivity preferred more over other biosensors. These biosensors can be further divided into potentiometric, conductometric, and amperometric [17, 18, 19]. Amperometric biosensors are enzyme-based sensors, extensively studied worldwide, and currently, the most common biosensors are available in the consumer market. These biosensors detect the direct or indirect current generated due to the electron exchange between the biological system and the electrode of the device.

The measurement of glucose level is done by using three enzymes that participate in the glucose metabolism pathways those are glucose-1-dehydrogenase (GDH), hexokinase, and glucose oxidase (GOx) [20– 21]. The basic working principle of glucose biosensors is based on the biological reaction, in which β-D-glucose is oxidized by gluconic acid and hydrogen peroxide, catalyzed by the GOx [22]. In this reaction, flavin adenine dinucleotide (FAD) acts as a co-factor and helps GOx to catalyze the reaction. FAD accepts the electron and is reduced to $FADH_2$.

$$Glucose + GOx\text{ –}FAD^+ \rightarrow Glucolactone + GOx\text{ –}FADH_2$$

The co-factor, *i.e.*, FAD, is generated due to the reaction with oxygen, which causes the libration of hydrogen peroxide also.

$$GOx\text{-}FADH_2 + O_2 \rightarrow GOx\text{-}FAD + H_2O_2$$

The electrons are generated due to the oxidation of H_2O_2 at the platinum electrode, and these electron counts are directly proportional to the glucose molecule present in the body's blood [23].

Body Fluids that can be used to Monitor the Glucose Level

As we know, most biosensors use blood in Fig. (**2**) to monitor glucose levels, but it causes pain, however, it's more accurate than any other method. Scientists have used other ways to monitor the glucose level by using body fluids such as urine, sweat, saliva, tear, *etc.* Some of them are discussed below in detail.

Urine

Urine is produced naturally from the body due to digestion and metabolism, which is easy to use for the analysis of glucose levels. It does not require any implant or forced extraction of fluid. So, this fluid of the body is more practical, feasible, and non-invasive for glucose level detection in the body. The glucose level can be measured by observing any changes in the urine refractive index in terms of glucose concentration [24]. It can also be measured by recording the resonance frequency of urine because it is directly proportional to the glucose concentration [25].

A photonic crystal sensor is used to detect any changes in the refractive index of urine. This sensor is two-dimensional and based on the resonator detector, which measures the output of urine refractive index changes [24]. The most valuable feature of any sensor is the speed of the resulting output, and this sensor fits best in this category. It is just 11.4 x 11.4 μm in size and gives the result in a few seconds, which is very important for a patient. This sensor not only predicts the result of urine glucose level but also can determine the bilirubin, blood glucose level, and albumin concentration in the patient body.

Saliva

The concept of using saliva to detect the body's glucose level was first introduced in a system that contained the bienzyme and carbon nanotubes (CNT) [26]. This device worked by using the horseradish peroxidase and GOx embedded in a CNT that reacts with bienzyme, entrapped in a chitosan matrix. Polyphenol is also used as a semi-permeable membrane that stops factors that can interfere with the accuracy of glucose level monitoring. The phosphate buffer solution (PBS) provided a reaction condition and tried to prevent the pH change of the reaction medium. Sodium chloride (NaCl) was tested as a factor that can help in stimulating saliva in more than 30 people. The major factors that interfere in glucose level measurement through saliva are salivary proteins and inhibitory biomolecules that need to be managed by the sensors [26– 27]. Another major problem with these biosensors is the precision and sensitivity against the glucose concentration because these are not that sensitive, which needs to be developed before mass testing [28].

Sweat

Sweat can be very useful for monitoring glucose levels in the human body. The first attempt at using sweat as a medium for glucose measurement used gold-coated nanoparticles in a single-layered graphene biosensor. The nanoparticles contain GOx, which utilizes the electrons generated by the carbon electrode and six-hexane-thiol [29]. A low-cost biosensor based on the gold and platinum nanoparticles in which GOx and chitosan are embedded onto the surface was introduced [30]. This device is cheap, highly sensitive, and simple in design.

Principles of Non-Invasive Glucose Monitoring

Multitude principles and techniques of non-invasive glucose monitoring have been pursued in academic research as well as in industry in recent times. Among them, four non-invasive glucose monitoring principles are widely investigated and reported. These four principles are differentiated based on their principles of detection of glucose. They are as follows:

• Optical Spectroscopy (Optical Detection)
• Photoacoustic Spectroscopy (Acoustic Detection)
• Electromagnetic Sensing (Electromagnetic Detection)
• Nanomaterial Based Sensing (Electrochemical Detection)

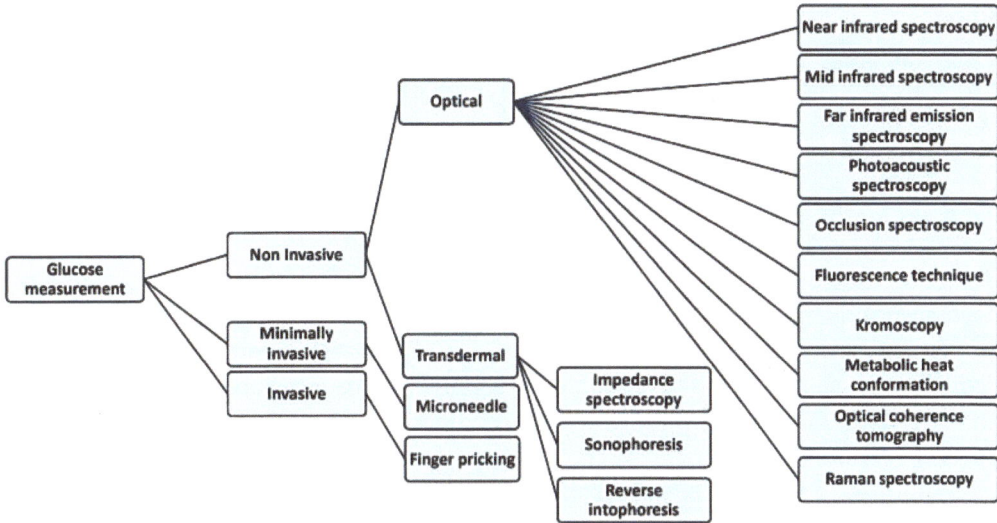

Fig. 2. Different classifications of blood glucose measurement.

Optical Spectroscopy

The basis for optical spectroscopy's detection principle is that glucose modifies optical signals by absorbing light at specific wavelengths. In the mid-IR and near-IR spectrum areas, overtones and combination bands are wavelengths. Due to this absorption at specific wavelengths, light transmission through glucose, both *in-vitro* and *in-vivo*, reduces when glucose concentration rises and *vice versa*. This phenomenon is used to quantify the blood glucose levels from the intensity of light that is transmitted or reflected using optical sensors like photodiodes [31]. Surface Plasmon Resonance (SPR), in which a beam of light is sent through the prism on the back of the SPR metal surface and bends onto the detector, is another technique for measuring glucose levels *via* optical detection. In contrast to light intensity, the approach uses data on angular shift to determine the glucose concentration. The light excitation of the electrons on the metal portion of the chip occurs at a specific resonance angle (refractive index). The shift in the reflection intensity curve is seen when an analyte is added to the SPR surface. The change in the angle of the SPR reflection intensity curve is used as the basis for the direct measurement of blood sugar [32].

Photoacoustic Spectroscopy

To determine glucose levels, photoacoustic glucose monitoring uses a hybrid technique that combines optical excitation with acoustic detection. Here, a multistage energy conversion process transforms the optical energy from the stimulation into acoustic energy [33]. Blood glucose molecules are optically excited, which causes the solution to warm up locally. This tiny temperature increase causes the optical interaction zone to expand thermally in volume. An ultrasonic piezoelectric transducer then measures the corresponding ultrasonic pressure pulse, which is then translated into electrical impulses to determine the glucose content.

Electromagnetic Sensing

Since the first decades of the 20th century, researchers have studied how body tissue responds to frequency. The permittivity of a given medium determines how electromagnetic waves will travel through it. Each material has a certain permittivity, a frequency-dependent property. This research has revealed a substantial frequency dependence in the complicated permittivity of tissue. Debye's relaxation theory serves as the foundation for the underlying mathematical idea. The equations can be expressed as follows [34– 35]:

$$\epsilon_r(\omega, \chi) = \epsilon_\infty(\chi) + \frac{\epsilon_{stat}(\chi) - \epsilon_\infty(\chi)}{1 + j\omega\tau(\chi)}$$

where, $\epsilon_r(\omega)$ is the complex and frequency-dependent relative permittivity of a dispersive material, ϵ_{stat} is the static permittivity at lower frequencies, ϵ_∞ is the permittivity at high frequencies and τ is the characteristic relaxation time of the medium. ω and χ are the angular frequency and the concentration, respectively. Plasma accounts for 55% of human blood. 90% or more of the blood plasma is water. Water has a high degree of polarization due to its dipolar nature. Due to this, the relative permittivity is also high. Glucose, on the other hand, is less polarized and has a smaller relative permittivity.

Thus, the relative permittivity of blood plasma is decreased or increased depending on whether the overall glucose concentration increases or decreases in the same volume of the blood sample. The measurement of blood glucose is aided by the measurement of the relative permittivity using a variety of electromagnetic sensing techniques at a certain microwave/mm-wave frequency, which provides the value of the blood glucose concentration as measured by electrical power. In other words, the electromagnetic interaction between the sensor and the material under test (MUT) - either blood or glucose solutions in a cuvette - where the amplitude and/or phase variations in the scattering parameters are measured, when the dielectric properties of the MUT change - represents the common measurement principle of these sensors. It is significant to highlight those variations in blood glucose levels have a far bigger impact on the dielectric properties of blood than do alterations in other blood constituents [36– 37].

Nanomaterial-Based Sensing

The development of nanomaterials as primary elements in sensing technologies, ranging from metallic gold, silver, copper oxide, and iron oxide to polymer composites, carbon nanotubes, and graphene, has significantly improved contemporary biosensors by obtaining valuable physiological data and useful information from vital body fluids like urine, saliva, sweat, and tears. Nanomaterials have a tremendous impact on sensing applications because they have huge surface areas, greater sensitivity and selectivity, and improved catalytic activities - all of which are necessary for a precise and accurate prediction of blood glucose levels in humans [38, 39, 40].

CONCLUSION

Diabetes is becoming a silent pandemic, which requires diagnosing correctly. If it's uncontrolled or untreated, then diabetes can cause the death of the patient. The biochemical test is expensive because it requires many chemicals, instruments, and working hands and is time-consuming, whereas the biosensors are quick, easy to handle, low cost, and accurate. These biosensors are used to monitor the glucose level from many biological fluids of the patient, like blood, sweat, *etc.* However, improvements are required in these biosensors so that they can predict much more accurately glucose levels from very less body fluids.

ACKNOWLEDGEMENT

Declared none.

REFERENCES

[1]   S.K. Dubey, V.K. Chaturvedi, D. Mishra, A. Bajpeyee, A. Tiwari, and M.P. Singh, "Role of edible mushroom as a potent therapeutics for the diabetes and obesity", *3 Biotech,* vol. 9, pp. 1-12, 2019. [http://dx.doi.org/10.1007/s13205-019-1982-3]

[2]   "Diabetes Australia", *Diabetes globally*. Available at: https://www.diabetesaustralia.com.au/about-diabetes/diabetes-globally/ (Accessed on: 2021).

[3]   D. Bruen, C. Delaney, L. Florea, and D. Diamond, "Glucose sensing for diabetes monitoring: recent developments", *Sensors (Basel),* vol. 17, no. 8, p. 1866, 2017. [http://dx.doi.org/10.3390/s17081866] [PMID: 28805693]

[4]   "World Health Organization (WHO)", *Diabetes*. Available at: https://www.who.int/health-topics/diabetes#tab=tab_3 (Accessed on: 2021).

[5]   A.J. Dunkley, D.H. Bodicoat, C.J. Greaves, C. Russell, T. Yates, M.J. Davies, and K. Khunti, "Diabetes prevention in the real world: effectiveness of pragmatic lifestyle interventions for the prevention of type 2 diabetes and of the impact of adherence to guideline recommendations: a systematic review and meta-analysis", *Diabetes Care,* vol. 37, no. 4, pp. 922-933, 2014. [http://dx.doi.org/10.2337/dc13-2195] [PMID: 24652723]

[6]   E.M. Benjamin, "Self-Monitoring of Blood Glucose: The Basics", *Clin. Diabetes,* vol. 20, no. 1, pp. 45-47, 2002.

[http://dx.doi.org/10.2337/diaclin.20.1.45]

[7]   G. Aleppo, K.J. Ruedy, T.D. Riddlesworth, D.F. Kruger, A.L. Peters, I. Hirsch, R.M. Bergenstal, E. Toschi, A.J. Ahmann, V.N. Shah, M.R. Rickels, B.W. Bode, A. Philis-Tsimikas, R. Pop-Busui, H. Rodriguez, E. Eyth, A. Bhargava, C. Kollman, and R.W. Beck, "REPLACE-BG: A randomized trial comparing continuous glucose monitoring with and without routine blood glucose monitoring in adults with well-controlled type 1 diabetes", *Diabetes Care,* vol. 40, no. 4, pp. 538-545, 2017.
[http://dx.doi.org/10.2337/dc16-2482] [PMID: 28209654]

[8]   M. Lind, W. Polonsky, I.B. Hirsch, T. Heise, J. Bolinder, S. Dahlqvist, E. Schwarz, A.F. Ólafsdóttir, A. Frid, H. Wedel, E. Ahlén, T. Nyström, and J. Hellman, "Continuous glucose monitoring vs conventional therapy for glycemic control in adults with type 1 diabetes treated with multiple daily insulin injections", *JAMA,* vol. 317, no. 4, pp. 379-387, 2017.
[http://dx.doi.org/10.1001/jama.2016.19976] [PMID: 28118454]

[9]   J. Kim, A.S. Campbell, B.E.F. de Ávila, and J. Wang, "Wearable biosensors for healthcare monitoring", *Nat. Biotechnol.,* vol. 37, no. 4, pp. 389-406, 2019.
[http://dx.doi.org/10.1038/s41587-019-0045-y] [PMID: 30804534]

[10]   A.P.F. Turner, "Tech.Sight. Biochemistry. Biosensors--sense and sensitivity", *Science,* vol. 290, no. 5495, pp. 1315-1317, 2000.
[http://dx.doi.org/10.1126/science.290.5495.1315] [PMID: 11185408]

[11]   L.C. Clark Jr, and C. Lyons, "Electrode systems for continuous monitoring in cardiovascular surgery", *Ann. N. Y. Acad. Sci.,* vol. 102, no. 1, pp. 29-45, 1962.
[http://dx.doi.org/10.1111/j.1749-6632.1962.tb13623.x] [PMID: 14021529]

[12]   A. Hiratsuka, K. Fujisawa, and H. Muguruma, "Amperometric biosensor based on glucose dehydrogenase and plasma-polymerized thin films", *Anal. Sci.,* vol. 24, no. 4, pp. 483-486, 2008.
[http://dx.doi.org/10.2116/analsci.24.483] [PMID: 18403839]

[13]   S.J. Updike, and G.P. Hicks, "The enzyme electrode", *Nature,* vol. 214, no. 5092, pp. 986-988, 1967.
[http://dx.doi.org/10.1038/214986a0] [PMID: 6055414]

[14]   J.P. Chambers, B.P. Arulanandam, L.L. Matta, A. Weis, and J.J. Valdes, "Biosensor recognition elements", *Curr. Issues Mol. Biol.,* vol. 10, no. 1-2, pp. 1-12, 2008.
[PMID: 18525101]

[15]   S.S. Iqbal, M.W. Mayo, J.G. Bruno, B.V. Bronk, C.A. Batt, and J.P. Chambers, "A review of molecular recognition technologies for detection of biological threat agents", *Biosens. Bioelectron.,* vol. 15, no. 11-12, pp. 549-578, 2000.
[http://dx.doi.org/10.1016/S0956-5663(00)00108-1] [PMID: 11213217]

[16]   J.D. Newman, and A.P. Turner, "Biosensors: principles and practice", *Essays Biochem.,* vol. 27, pp. 147-159, 1992.
[PMID: 1425600]

[17]   K. Habermüller, M. Mosbach, and W. Schuhmann, "Electron-transfer mechanisms in amperometric biosensors", *Fresenius J. Anal. Chem.,* vol. 366, no. 6-7, pp. 560-568, 2000.
[http://dx.doi.org/10.1007/s002160051551] [PMID: 11225768]

[18]   J.E. Pearson, A. Gill, and P. Vadgama, "Analytical aspects of biosensors", *Ann. Clin. Biochem.,* vol. 37, no. 2, pp. 119-145, 2000.
[http://dx.doi.org/10.1258/0004563001899131] [PMID: 10735356]

[19]   D.R. Thévenot, K. Toth, R.A. Durst, and G.S. Wilson, "Electrochemical biosensors: recommended definitions and classification", *Biosens. Bioelectron.,* vol. 16, no. 1-2, pp. 121-131, 2001.
[PMID: 11261847]

[20]   C.P. Price, "Point-of-care testing in diabetes mellitus", *Clin. Chem. Lab. Med.,* vol. 41, no. 9, pp. 1213-1219, 2003.
[http://dx.doi.org/10.1515/CCLM.2003.185] [PMID: 14598871]

[21]   E.J. D'Costa, I.J. Higgins, and A.P.F. Turner, "Quinoprotein glucose dehydrogenase and its application in an amperometric glucose sensor", *Biosensors,* vol. 2, no. 2, pp. 71-87, 1986.
[http://dx.doi.org/10.1016/0265-928X(86)80011-6] [PMID: 3454651]

[22]   M.K. Weibel, and H.J. Bright, "The glucose oxidase mechanism. Interpretation of the pH dependence", *J. Biol. Chem.,* vol. 246, no. 9, pp. 2734-2744, 1971.
[http://dx.doi.org/10.1016/S0021-9258(18)62246-X] [PMID: 4324339]

[23]   G.G. Guilbault, and G.J. Lubrano, "An enzyme electrode for the amperometric determination of glucose", *Anal. Chim. Acta,* vol. 64, no. 3, pp. 439-455, 1973.
[http://dx.doi.org/10.1016/S0003-2670(01)82476-4] [PMID: 4701057]

[24]   S. Robinson, and N. Dhanlaksmi, "Photonic crystal based biosensor for the detection of glucose concentration in urine", *Photonic Sens.,* vol. 7, no. 1, pp. 11-19, 2017.
[http://dx.doi.org/10.1007/s13320-016-0347-3]

[25]   G. Koirala, R. Dhakal, E.S. Kim, Z. Yao, and N.Y. Kim, "Radio frequency detection and characterization of water-ethanol solution through spiral-coupled passive micro-resonator sensor", *Sensors (Basel),* vol. 18, no. 4, p. 1075, 2018.
[http://dx.doi.org/10.3390/s18041075] [PMID: 29614033]

[26]   J. Liu, S. Sun, H. Shang, J. Lai, and L. Zhang, "Electrochemical biosensor based on bienzyme and carbon nanotubes incorporated into an os-complex thin film for continuous glucose detection in human saliva", *Electroanalysis,* vol. 28, no. 9, pp. 2016-2021, 2016.
[http://dx.doi.org/10.1002/elan.201501179]

[27]   T. Arakawa, K. Tomoto, H. Nitta, K. Toma, S. Takeuchi, T. Sekita, S. Minakuchi, and K. Mitsubayashi, "A wearable cellulose acetate-coated mouthguard biosensor for *In Vivo* salivary glucose measurement", *Anal. Chem.,* vol. 92, no. 18, pp. 12201-12207, 2020.
[http://dx.doi.org/10.1021/acs.analchem.0c01201] [PMID: 32927955]

[28]   K. Iwai, T. Minamisawa, K. Suga, Y. Yajima, and K. Shiba, "Isolation of human salivary extracellular vesicles by iodixanol density gradient ultracentrifugation and their characterizations", *J. Extracell. Vesicles,* vol. 5, no. 1, p. 30829, 2016.
[http://dx.doi.org/10.3402/jev.v5.30829] [PMID: 27193612]

[29]   Y. Yuan, Y. Wang, H. Wang, and S. Hou, "Gold nanoparticles decorated on single layer graphene applied for electrochemical ultrasensitive glucose biosensor", *J. Electroanal. Chem. (Lausanne),* vol. 855, no. 1, p. 113495, 2019.
[http://dx.doi.org/10.1016/j.jelechem.2019.113495]

[30]   X. Xuan, H.S. Yoon, and J.Y. Park, "A wearable electrochemical glucose sensor based on simple and low-cost fabrication supported micro-patterned reduced graphene oxide nanocomposite electrode on flexible substrate", *Biosens. Bioelectron.,* vol. 109, no. 1, pp. 75-82, 2018.
[http://dx.doi.org/10.1016/j.bios.2018.02.054] [PMID: 29529511]

[31]   O.S. Khalil, "Spectroscopic and clinical aspects of noninvasive glucose measurements", *Clin. Chem.,* vol. 45, no. 2, pp. 165-177, 1999.
[http://dx.doi.org/10.1093/clinchem/45.2.165] [PMID: 9931037]

[32]   A. Omidniaee, S. Karimi, and A. Farmani, "Surface plasmon resonance-based SiO$_2$ kretschmann configuration biosensor for the detection of blood glucose", *Silicon,* vol. 14, no. 6, pp. 3081-3090, 2022.
[http://dx.doi.org/10.1007/s12633-021-01081-9]

[33]   H.A. MacKenzie, H.S. Ashton, S. Spiers, Y. Shen, S.S. Freeborn, J. Hannigan, J. Lindberg, and P. Rae, "Advances in photoacoustic noninvasive glucose testing", *Clin. Chem.,* vol. 45, no. 9, pp. 1587-1595, 1999.
[http://dx.doi.org/10.1093/clinchem/45.9.1587] [PMID: 10471673]

[34]   P.J.W. Debye, "Polar molecules", In: *Chemical Catalog Company*, 1929.

[35]   K.S. Cole, and R.H. Cole, "Dispersion and absorption in dielectrics I. Alternating Current Characteristics", *J. Chem. Phys.,* vol. 9, no. 4, pp. 341-351, 1941.
[http://dx.doi.org/10.1063/1.1750906]

[36]   N.Y. Kim, K.K. Adhikari, R. Dhakal, Z. Chuluunbaatar, C. Wang, and E.S. Kim, "Rapid, sensitive, and reusable detection of glucose by a robust radiofrequency integrated passive device biosensor chip", *Sci. Rep.,* vol. 5, no. 1, p. 7807, 2015.
[http://dx.doi.org/10.1038/srep07807] [PMID: 25588958]

[37]   K.K. Adhikari, and N-Y. Kim, "Ultrahigh-sensitivity mediator-dree biosensor based on a microfabricated microwave resonator for the detection of micromolar glucose concentrations", *IEEE Trans. Microw. Theory Tech.,* vol. 64, no. 1, pp. 319-327, 2016.
[http://dx.doi.org/10.1109/TMTT.2015.2503275]

[38]   P. Makaram, D. Owens, and J. Aceros, "Trends in nanomaterial-based non-invasive diabetes sensing technologies", *Diagnostics (Basel).,* vol. 4, no. 2, pp. 27-46, 2016.
[http://dx.doi.org/ 10.3390/diagnostics4020027] [PMID: 26852676]

[39]   V.K. Chaturvedi, S.K. Dubey, and M.P. Singh, "Antidiabetic potential of medicinal mushrooms.Phytochemicals from medicinal plants", In: *Phytochemicals from medicinal plants* Apple Academic Press, 2019, pp. 137-158.
[http://dx.doi.org/10.1201/9780429203220-7]

[40]   M.M. Rahman, A.J.S. Ahammad, J.H. Jin, S.J. Ahn, and J.J. Lee, "A comprehensive review of glucose biosensors based on nanostructured metal-oxides", *Sensors (Basel),* vol. 10, no. 5, pp. 4855-4886, 2010.
[http://dx.doi.org/10.3390/s100504855] [PMID: 22399911]

CHAPTER 7

Role of Paper-based Biosensor in Diagnostics

Sujeet Singh[1], **Praveen Rai**[2], **Hemant Arya**[1] and **Vivek K. Chaturvedi**[3,*]

[1] *Department of Biotechnology, Central University of Rajasthan, Ajmer, Rajasthan 305817, India*

[2] *National Institute of Plant Genome Research, New Delhi, India*

[3] *Department of Gastroenterology, Institute of Medical Sciences, Banaras Hindu University, Varanasi, Uttar Pradesh, India*

Abstract: New diagnostic technologies are paper-based sensors that are multifunctional, highly flexible, absorbent, and environmentally friendly. The substrate can be used to design a cost-effective framework for disease detection, prognosis, and surveillance of illnesses that is easy, reliable, and quick in our medical healthcare sector. Paper-based devices are an extremely cheap innovation for fabricating simplified and movable diagnosing processes that can be extremely useful in resource-constrained settings like developing countries, where fully equipped infrastructure and highly skilled medical persons are unavailable. Point-of-care (POC) devices give a significant advantage over traditional procedures for *in-situ* measurement of illness or disease biomarkers, assisting physicians in making decisions. Paper-based analytical devices that combine paper substrates have become popular point-of-need diagnostics over the last decade. We discuss in this chapter the paper-based analytical biosensors and the classification of paper-based biosensors (PBBs) as Dipstick tests, lateral flow assay (LFAs), microfluidic biosensors, and biosensor devices (transducers and biorecognition elements). Furthermore, paper-based biosensors are used to detect malaria and other diseases.

Keywords: Biomarker, Biosensor, Diagnosis, Fabrication, Malaria, Paper, Printing.

INTRODUCTION

Leland C. Clerk discovered the first biosensor in 1956 to detect oxygen [1]. These analytical devices measure the concentration of analytes [2]. For example, in measuring sugar levels in diabetes patients, we use a glucometer, a type of biosensor. There are various types of biomolecules present inside the human cell. The biosensors detect the concentration of a specific biomolecule and analyze human health issues (Fig. **1**). The biosensor is a compact device that helps to

---

* **Corresponding author Vivek K. Chaturvedi:** Department of Gastroenterology, Institute of Medical Sciences, Banaras Hindu University, Varanasi, Uttar Pradesh, India; E-mail: vkchaturvedi.nbt@gmail.com

Vivek K. Chaturvedi, Dawesh P. Yadav and Mohan P. Singh (Eds.)

determine human health and disease conditions, and the measured biomolecules are called biomarkers [1].

The blood, urine, saliva, and sweat samples were taken from the human body for continuous measurement through biosensor devices. With the help of biosensor devices, patients and doctors immediately get the correct information. It also reduces unnecessary hospital visits and promotes self-management of health and diseases [1]. There are different biosensors, like DNA-based biosensors, piezoelectric biosensors, enzyme-based biosensors, tissue-based biosensors, thermal-based biosensors, and paper-based biosensors.

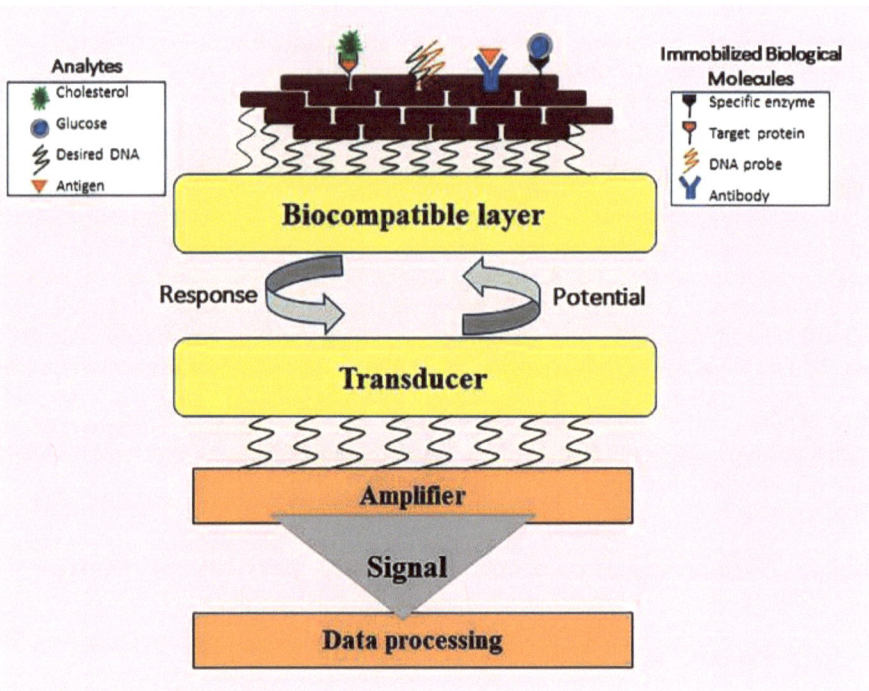

Fig. (1). Working principle of biosensors.

Paper-based Biosensors

To ensure adequate human healthcare, we want rapid, simple, accurate, and low-cost detection methods that can immediately identify the onset of human diseases by using the disease biomarkers in body fluids (serum, blood, urine, saliva, and sweat). Similarly, traditional ELISA (enzyme-linked immunosorbent assay) and spectrophotometric techniques are mostly used to quantify environmental contaminants (*e.g.*, heavy metal ions in industrial wastewater effluent). These biosensor devices did not require skilled personnel to perform diagnostic tests and evaluate results in resource-constrained areas. The use of paper-based biosensor devices comprises the detection of diseases, examining human health conditions, detecting of any biological pathogens, and detecting water and food quality (Fig. **2**). The traditional method for detecting any disease takes a long time and requires a sophisticated device. Still, with the help of point-of-care (POC) devices, it has opened a wide range of diagnoses for *in-situ* analyses. The paper-based biosensor devices have many advantages. The paper-based biosensor is highly flexible, eco-friendly, polymeric, cost-efficient, hugely available, less weighted, and hydrophilic. Due to the porous nature of the paper, the liquid solution flows *via* capillary action, and no external pump is required. In the sample, detecting the analyte due to its chemical reaction can cause a color change, electrochemical properties, and light emission or absorption [3]. The most common detection method depends on the colorimetric changes. The evaluation of the result is to identify the color product generated due to the binding of ligand-analyte, which scanners, cameras, or mobile phones can further quantify. The use of paper-based biosensors has many advantages. (i) it provides a high surface-to-volume ratio; (ii) compatibility with the biological samples; (iii) capillary action; (iv) adsorption properties; and (v) immobilization of antibodies and proteins. The incineration of paper-based biosensors is easy and quickly accessible globally [4]. After all, fabrication techniques for paper-based microfluidic devices, such as wax printing, are very cheap.

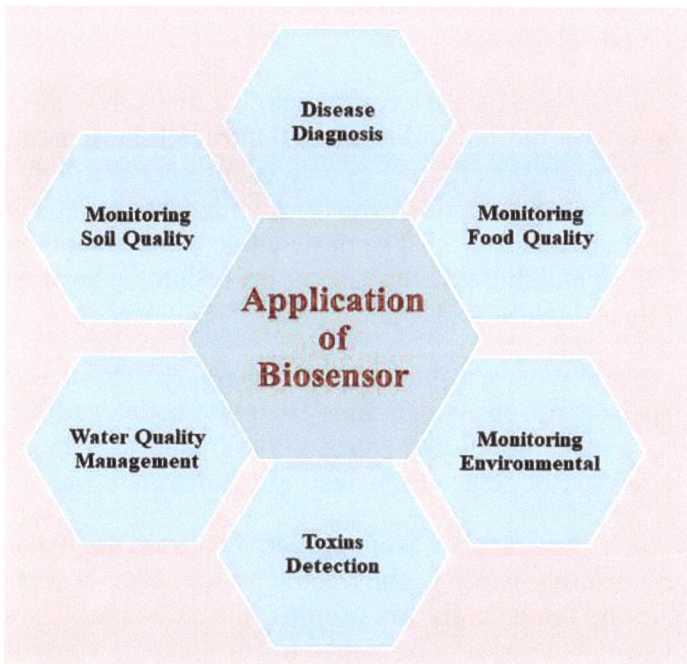

Fig. (2). Application of biosensors.

Even the use of paper-based sensors has some limitations when it comes to identifying multiple biomolecules at the same time. Furthermore, colorimetric methods such as electrochemiluminescence, fluorescence, and chemiluminescence, along with electrochemistry, are also utilized to monitor analytes. The processes are executed in the paper-based microfluidic program, but the methods also display a few drawbacks related to cost, sensitivity, and simplicity [5].

The current fabrication method varies from wax printing to kirigami, and origami-encouraged methodologies have unique advantages for the identified analytes. The paper-based sensors use various detection methods. There are numerous applications, ranging from disease diagnosis to environmental contaminants. Finally, through the discussion, we determined the possible future purposes of paper-based sensors.

Types of Paper

Due to its excellent wicking capabilities and alpha-cellulose composition, chromatography paper (Whatman brand) is the most commonly used [6 - 9]. Due to its thickness, the average retention and movement in a specific type of paper are 180 μm, and the size of pores is 11 μm. The new kind of paper, like Whatman filter paper number 4, was also utilized because of its bigger pore size (20–25) μm and high retention rate [10].

Biosensor devices may need a variety of chemical and physical characteristics. The immobilization of biomolecules in the nitrocellulose membrane due to functional groups permits the creation of a covalent link. Weak hydrogen bonds, ion interactions, and Vander Waals interactions with protein and substrates are all possible *via* the membrane [9 - 11]. The nitrocellulose membrane has a consistent pore size of 0.45 μm, and these membranes (nitrocellulose) have been altered by wax printing followed by heating [10].

A similar investigation was conducted on glossy paper, demonstrating the possibility of paper-based sensors. Organic filler and cellulose fiber are mixed in the paper matrix to make the glossy paper. Glossy paper is also used to create a flexible paper-based sensor device for detecting various chemical pollutants [12].

Various types of ordinary paper were effectively used to make paper-based sensors. Standard printing paper is chosen for future usage in wearable devices, which are flexible and foldable. In wax printing, the carbon ink stays dispersed on the entire surface of the paper. The paper towel makes a biosensor due to its wicking properties. Filter paper is more expensive than towel paper because of its high porosity, which makes it a very feasible material for analyzing various analytes [13].

IMPORTANCE OF PRINTING AND FABRICATION

The fabrication method is most widely used to create biosensors that can improve usability and functionality. Several ways are necessary for the physical or chemical sedimentation of the paper. Both methods affect the raw material properties of the cellulose component. In this book chapter, the various techniques will be discussed.

Wax Printing

The creation of microfluidic channels by using the wax printing method is a unique way to create a paper-based sensor. The wax printing method is used primarily for its minimal cost, simplicity, and non-toxicity compared to the other fabrication methods [14]. The wax pattern, which is directly printed on the paper, regulates the movement of the samples and reagent fluid. The wax was melted down on the chromatography paper using the hot plate devices to generate a hydrophobic barrier channel. Using a hair dryer to melt the wax material onto the paper [15]. In another study, wax microstructures were printed on the nitrocellulose membranes and baked to allow the print wax to penetrate and form hydrophobic barrier channels in the devices [11]. The wax printing procedure takes just 10–15 minutes. But wax printing is not appropriate for high-resolution printing. These patterns are not stable at high temperatures due to the melting nature of the wax.

Photolithography

Fabrication of a paper-based sensor device with high resolution and the ability to detect multiple biomarkers and large-scale production by the photolithography approach. After that, the paper is baked with dry heat. The paper is then coated with a patterned transparent film created by laser printing and irradiation with ultraviolet (UV) light. The unpolymerized photosensitivity was removed with acetone after additional dry heating, and the paper was subjected to air plasma to form hydrophilic regions. The photolithography process involves using costly equipment and includes lengthy and complicated procedures [16]. However, these paper-based sensors can be damaged once they are folded or bent.

Polydimethylsiloxane (PDMS) Printing

The printing liquid of PDMS sticks to the paper owing to its hydrophobic nature. Fabrication of flexible sensors through PDMS patterning on the paper [17]. The

PDMS patterned paper sensors do not damage the microfluidic channels when the sensor is folded or bent. In this fabrication method, the ratio of PDMS and hexanes 3:1 (PDMS: hexanes) proportion was inserted on a plotter and printed on top of the paper. The integrated double layer of PDMS is composed of a patterned layer and a barrier channel layer. To generate a pattern, PDMS flowed over the stencil, and then the channel layer and patterned layer were combined to produce the microfluidic channel [17].

Inkjet Printing

Inkjet printing has been permitted in the direction of simplifying the fabrication procedure by just depending on the purpose of their inkjet printing equipment, devoid of the necessity for the other equipment. The popularity of inkjet printing is due to the minimal cross-contamination between the fluid (sample and reagent), which is a non-contact operation. Typically, commercial printing devices remain utilized, either with or without any alteration to the way they print the biological molecules designed for the sensors. The ink cartridges and the Canon inkjet printer are altered to fabricate paper-based microfluidic devices [18, 19]. However, the bio-ink printer is costly, and inkjet printing devices must have been changed.

Laser Cutting

Laser cutting is a straightforward and affordable approach to creating designs on paper-based devices. Chitnis *et al.* used a CO_2 laser cutter to create hydrophilic patterns on paper, which were then changed with a hydrophobic coat. Using a laser-cutting approach, this group successfully converted hydrophobic zones on the paper's surface and inside into hydrophilic parts. The method generates hydroxyl and carbonyl groups on the paper surface [20]. The laser cutter cannot cut across the paper completely. A CO_2 laser was mainly used to form barrier channels by shedding the paper's hydrophilic fabric. Although the laser cutting process is simple, it requires precision tools for the laser cutter, engraving, 2D graphics creation software, and both-sided adhesive to produce paper devices.

Hot Embossing

Hot embossing is a procedure for fabricating paper-based sensors to create hydrophilic channel barriers that remain in the cavity, which permits spontaneous capillary movement. Hot embossing is an effective process, as the spin-to-spin of the hot embossing method allows for high quantities due to its low cycle times. The fabrication of a microfluidic device with the hot embossing PowerCoat[R] HD paper (Boulogne-Billancourt, Paris, France) coated with a layer of rubber for water resistance and polyvinyl alcohol for hydrophilicity [21]. However, hot embossing can be efficient, and it also requires the use of specific instruments.

Hydrophobic Silanization

Hydrophobic Silanization is a method to cover the substrate's surface along with the alkoxysilane molecules. Yang *et al.* also displayed that silanizing paper and trichloromethylsilane, facilitate the formation of hydrophobic substrates [22]. The hydrophobic silanization method remains easy, inexpensive, and quickly performed. Filter paper is silanized by soaking it in 2% trimethoxy (octadecyl) silane and heating it for one hour. Subsequently, the hydrophobic paper was associated with the paper mask in print with an easily accessible laser printer. Wet etching with sodium hydroxide (NaOH) is used for saturated hydrophilic regions [23]. Hydrophobic silanization is a low-cost procedure that does not require a piece of costly equipment; the fabrication takes about 7-8 minutes.

Kirigami and Origami

The purpose of kirigami, which means cutting paper, and origami, which means folding paper, are mainly used to fabricate microfluidic devices. These methods have offered researchers new opportunities to fabricate their devices. The principles of origami were used to create the unique device in which a fold separates the two sections [8, 24]. The enzyme immobilization area is one, while the detection area is the other. Using the origami-inspired gadget, the researcher creates a paper-based analytical instrument to detect glucose electrochemically. In addition to origami and kirigami, paper-based gadgets with the 2D and 3D vertical movement were created [22]. Kirigami was mainly used to create different designs from a single cut-down set in which the paper is bent and then unbent.

DETECTION TECHNIQUE

The different types of techniques, fluorescence, colorimetric assays, electrochemical, electrochemiluminescence, and chemiluminescence, are based on methods that can be used in the paper-based sensor to detect the analyte of interest. These paper-based sensors technology have been highly used because they give susceptible results when evaluated to conventional techniques like ELISA [25].

Fluorescence

Fluorometry is the device that is used to determine the quantification of luminescence. Fluorescence is the technique that identifies the luminescence from fluorophores. It is because of the contact between the sample molecules [26]. One research group constructed a paper-based device that uses the same principle of fluorescence to help detect hazardous compounds like hydrogen sulfide (H_2S) [27]. A chemical reaction occurs between FMA and H_2S. The results are noted at the 470 nm wavelength. The response was generated in 60 seconds, and the limit of detection of H_2S was at three parts per billion (3 PPB). The fluorescent dye remains compatible with biological molecules. They were inspected on a continual basis, however, because they were subjected to a photobleaching process that damaged their fluorescence intensity.

Colorimetric Assays

Colorimetric assays include techniques for estimating color variation to determine the presence and amount of an analyte. a) direct image using a single lens reflex (SLR) camera, a cell phone camera, or the small-budget desktop scanners in the grouping of software like MATLAB for quantity estimation [7] or b) a conventional spectrophotometer device used to measure the absorbance of a particular wavelength. This method is widely used because it produces very precise results.

Electrochemical

The electrochemical sensor comprises three electrodes: the working electrode (WE), the reference electrode (RE), and the counter electrode (CE) [28 - 33]. The detection of the sample is done when the WE and CE link with the solution of electrolytes, which provides a flow of current between the WE and the model [5, 30, 31, 34]. These electrodes can be screen-printed on the surface chromatography paper using silver or silver chloride (Ag/AgCl) and carbon ink [32, 33, 35]. Recently, graphite pencils were used as another source for fabricating electrodes.

Electrochemiluminescence

In the electrochemiluminescence method, electrochemical reactions occur and produce luminescence. The reagent that helps in the production of luminescence is oxidized at WE (the screen-printed carbon) together with the hydrogen peroxide, or else triethylamine (TEA) is attained by the reduction of a sample [36]. The photons are generated due to the reaction, and the luminescent color difference occurs. The image assessment software (*e.g.*, NIH Image J and Python) was used to study the pixel intensity used for every color, specifying the amount of analyte concentration.

Chemiluminescence

The light appears due to a chemical reaction, and the chemiluminescence comprises the detection of this light; the reactive intermediate molecules cause emissions after reverting to the ground state from their excited state. Detecting the uric acid by chemiluminescence occurs when the enzymatic reaction helps produce hydrogen peroxide as a by-product by disintegrating the substrate [37]. The hydrogen peroxide reacts with the rhodamine derivative, which was added to the surface of the microfluidic channels of the sensor. Chemiluminescence is a very accurate and sensitive detection method. It is also considered one of the easiest detection methods because there is no need for highly sophisticated equipment [38]. Chemiluminescence still requires an enzyme, which is costly, prohibiting the paper's porous structure and affecting the detecting method's total efficiency.

PAPER-BASED SENSORS APPLICATION IN MEDICAL DIAGNOSTICS

There are many applications of paper-based sensors, such as disease diagnosis, detecting heavy metals in polluted water, detecting toxins, and checking food quality. The paper-based sensor also permits the rapid detection of diseases in underprivileged areas. These biosensor devices are alternatives to well-sophisticated laboratory tests and have been widely studied. A biosensor assay with more selectivity and high sensitivity is immediately needed to detect disease at a very early stage [39, 40]. The diagnosis of any disease uses various biological fluids such as invasive (blood) and non-invasive samples (tears, urine, and sweat) in paper-based biosensors.

THE USE OF BIOMARKERS FOR THE DETECTION OF MALARIA

Malaria is one of the most life-threatening diseases and is transmitted through the bite of female Anopheles mosquitoes infected with *Plasmodium* parasites. It creates a primary worldwide health concern in the tropical and subtropical zones. The malaria mortality rate is extremely high when compared with other protozoan-based diseases. According to the World Health Organization (WHO) 2021 data, there were 247 million disease cases and 6,19,000 deaths from malaria in 2021 (WHO Report 2021). The apicomplexan protozoan parasite *Plasmodium* is a single-cell organism. Mainly, five different *Plasmodium* parasite species are responsible for causing malaria disease in humans. They are *Plasmodium falciparum, P. vivax, P. ovale, P. malariae,* and *P. knowlesi.* Malaria completes its life cycle in two different hosts: female Anopheles mosquitos and humans.

The WHO aims to eradicate malaria by 2030 [41, 42]. This goal is only attained when all the malaria cases are ideally diagnosed with a species-specific approach which can help the guide to decide and the treatment is done correctly [43]. For malaria detection, various sensitive biosensors were used, including the enzymatic method, the non-enzymatic method, and the label-free method [44] (Fig. **3**). The most used rapid diagnostic tests (RDTs) are used to diagnose malaria, which is based on *P. falciparum* histidine-rich protein-2 (*Pf*HRP-2), which is specifically expressed in the *P. falciparum* parasite [45]. Due to the lack of the *Pf*HRP2 gene in several *P. falciparum* isolates, detection of malaria by *Pf*HRP-2 RDTs give a false negative result [46].

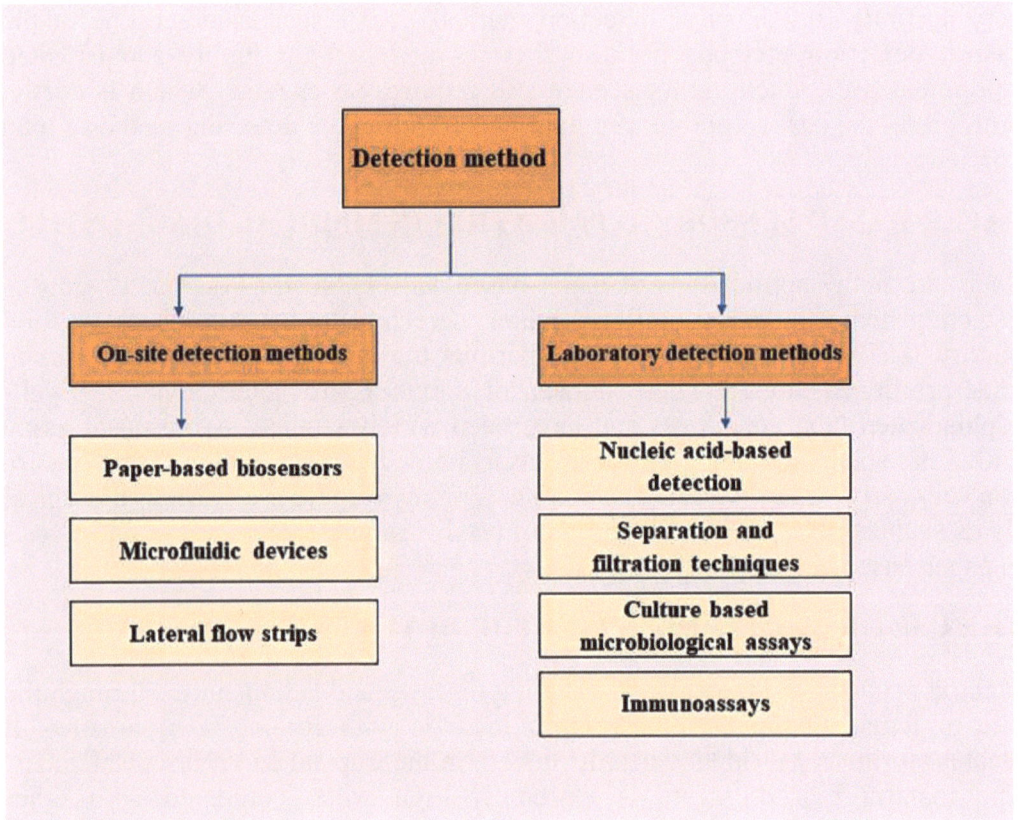

Fig. (3). Detection methods.

Hence, the new RDTs for malaria are based on the *Plasmodium* lactate dehydrogenase enzyme, expressed in all the *Plasmodium* species (*PLDH*). The *PLDH* biomarker is used to identify malaria since it is substantially expressed [41, 47]. The current approach for malaria detection uses a rapid diagnostic test based on lateral flow test approaches, in which the use of body fluids (blood, urine, and saliva) on the sensor strip allows it to bind with the target. The different zones are present on the biosensor strip and labeled or captured at the target site through biorecognition. The specific antibodies mainly interact with biomarkers to generate a unique signal (Malaria rapid diagnostic test performance). Results of WHO product testing of malaria RDTs: Round 8 (2016–2018). The designing of the microfluidic paper analytical device (μPAD) using the CorelDRAW software and the use of Whatman chromatography paper for printing using a thermal solid ink printer device (Xerox Coloqube 8870) created by the Council for Scientific and Industrial Research (CSIR) Microfluidic facility, South Africa [45]. The fabrication on the paper *via* a wax-printed channel specifically attaches the recombinant *P. falciparum* lactate dehydrogenase (rPfLDH) attachment with the aptamers on the sites to construct the microfluidic paper-based analytical device (μPAD). This type of design was commonly used to screen various rPfLDH-binding aptamers (like LDHH4, pL1, rLDH4, LDHp11, and 2008) for the capability to capture rPfLDH selectively [45].

The commercial approach to nano-biosensing technology and the signal amplification approaches have been assessed to formulate the commercially available method for malaria detection programs. Currently, several commercial programs have emerged [41, 48]. Many experiments are designed on portable malaria diagnosis microscopes and applied to a cell phone-based biosensor. The x-rapid system is now made saleable for the malaria diagnosis system for smartphone gadgets microscopy, which has the LED and the ten times lens system attached to the smartphone gadgets [49]. Nowadays, digital visualization for the malaria diagnosis setup might be used for bedside detection [50].

CONCLUSION

Currently, paper-based biosensors signify an evolving approach with many applications, such as monitoring forensic samples, disease diagnosis, and the detection of environmental agents. The paper-based biosensors are portable, low-cost devices that detect highly sensitive analytes. The paper-based biosensor's fabrication cost is too low, and the biosensor is beneficial not only in a fully equipped research laboratory but also in resource-constrained settings and adverse conditions. There are many advantages of paper-based biosensors: (a) no need for any equipment (no need for a pump for the movement of fluid); (b) easy to carry by nature; (c) straightforward and user-friendly; (d) disposal is straightforward; (e) need only a small number of analytes (microliter range); and (f) multiplexed analysis of analytes (the detection of more than one analyte). Also, the biosensing methods described in the book chapter, like color variation, give easy and straightforward detection; there is no need for highly skilled medical professionals to evaluate the results of the diagnostic disease tests. However, it is also essential to highlight some of the limitations of paper-based sensors that must be addressed before they can be commercialized. The paper-based sensors are subject to the sensitivity and accuracy of the analyte detection. Paper-based biosensors also have an application component for detecting disease biomarkers such as malaria. The economical and easy-to-carry POC devices use the remote area to monitor food and water quality. The paper-based sensor is also used to detect toxins and heavy metals in environmental samples like polluted water. Paper microfluidics not only make an easy diagnostic device, but they also enable the technology for the biomedical research area.

ACKNOWLEDGEMENTS

The author (Sujeet Singh) would like to thank the Central University of Rajasthan, Ajmer, Govt.of India, for providing financial assistance. V.K.C. gratefully acknowledges the Department of Health Research (DHR), Govt. of India, for support through the Young Scientist Fellowship Grant R.12014/56/2022-HR.

REFERENCES

[1]   S. Patel, R. Nanda, S. Sahoo, and E. Mohapatra, "Biosensors in health care: the milestones achieved in their development towards lab-on-chip-analysis", *Biochem Res Int.,* vol. 2016, p. 3130469, 2016.
[http://dx.doi.org/10.1155/2016/3130469] [PMID: 27042353]

[2]   O. Amor-Gutiérrez, E. Costa-Rama, and M.T. Fernández-Abedul, "Sampling and multiplexing in lab-on-paper bioelectroanalytical devices for glucose determination", *Biosens. Bioelectron.,* vol. 135, pp. 64-70, 2019.
[http://dx.doi.org/10.1016/j.bios.2019.04.006] [PMID: 30999242]

[3]   O.S. Oliinyk, A.A. Shemetov, S. Pletnev, D.M. Shcherbakova, and V.V. Verkhusha, "Smallest near-infrared fluorescent protein evolved from cyanobacteriochrome as versatile tag for spectral multiplexing", *Nat. Commun.,* vol. 10, no. 1, p. 279, 2019.
[http://dx.doi.org/10.1038/s41467-018-08050-8] [PMID: 30655515]

[4]   E. W. Nery, and L. T. Kubota, "Sensing approaches on paper-based devices: a review", *Analytical and Bioanalytical Chemistry,* vol. 405, pp. 7573-7595, 2013.
[http://dx.doi.org/10.1007/s00216-013-6911-4]

[5]   D.D. Liana, B. Raguse, J.J. Gooding, and E. Chow, "Recent advances in paper-based sensors", *Sensors (Basel),* vol. 12, no. 9, pp. 11505-11526, 2012.
[http://dx.doi.org/10.3390/s120911505] [PMID: 23112667]

[6]   A. W. Martinez, S. T. Phillips, E. Carillho, and G. M. Whitesides, "Diagnostics for the developing world: microfluidic paper-based analytical devices", *(in eng), Anal. Chem.,* vol. 82, pp. 3-10, 2010.
[http://dx.doi.org/10.1021/ac9013989]

[7]   A.K. Ellerbee, S.T. Phillips, A.C. Siegel, K.A. Mirica, A.W. Martinez, P. Striehl, N. Jain, M. Prentiss, and G.M. Whitesides, "Quantifying colorimetric assays in paper-based microfluidic devices by measuring the transmission of light through paper", *Anal. Chem.,* vol. 81, no. 20, pp. 8447-8452, 2009.
[http://dx.doi.org/10.1021/ac901307q] [PMID: 19722495]

[8]   B. Gao, J. Chi, H. Liu, and Z. Gu, "Vertical paper analytical devices fabricated using the principles of quilling and kirigami", *Sci. Rep.,* vol. 7, no. 1, p. 7255, 2017.
[http://dx.doi.org/10.1038/s41598-017-07267-9] [PMID: 28775253]

[9]   M. Cretich, V. Sedini, F. Damin, M. Pelliccia, L. Sola, and M.J.A.b. Chiari, "Coating of nitrocellulose for colorimetric DNA microarrays", *Anal Biochem.,* vol. 397, no. 1, pp. 84-88, 2010.
[http://dx.doi.org/10.1016/j.ab.2009.09.050] [PMID: 19800859]

[10] M.M. Ali, C.L. Brown, S. Jahanshahi-Anbuhi, B. Kannan, Y. Li, C.D.M. Filipe, and J.D. Brennan, "A printed multicomponent paper sensor for bacterial detection", *Sci. Rep.,* vol. 7, no. 1, p. 12335, 2017.
[http://dx.doi.org/10.1038/s41598-017-12549-3] [PMID: 28951563]

[11] Y. Lu, W. Shi, J. Qin, and B. Lin, "Fabrication and characterization of paper-based microfluidics prepared in nitrocellulose membrane by wax printing", *Anal. Chem.,* vol. 82, no. 1, pp. 329-335, 2010.
[http://dx.doi.org/10.1021/ac9020193] [PMID: 20000582]

[12] A. Arena, N. Donato, G. Saitta, A. Bonavita, G. Rizzo, and G. Neri, "Flexible ethanol sensors on glossy paper substrates operating at room temperature", *Sensors and Actuators B: Chemical.,* vol. 145, no. 1, pp. 488-494, 2010.
[http://dx.doi.org/10.1016/j.snb.2009.12.053]

[13] S. Cinti, V. Mazzaracchio, I. Cacciotti, D. Moscone, and F. Arduini, "Carbon black-modified electrodes screen-printed onto paper towel, waxed paper and parafilm M®", *Sensors (Basel),* vol. 17, no. 10, p. 2267, 2017.
[http://dx.doi.org/10.3390/s17102267] [PMID: 28972566]

[14] W. Dungchai, O. Chailapakul, and C.S. Henry, "A low-cost, simple, and rapid fabrication method for paper-based microfluidics using wax screen-printing", *Analyst (Lond.),* vol. 136, no. 1, pp. 77-82, 2011.
[http://dx.doi.org/10.1039/C0AN00406E] [PMID: 20871884]

[15] P. Namwong, P. Jarujamrus, M. Amatatongchai, and S.J.J.C.E. Chairam, "Fabricating simple wax screen-printing paper-based analytical sevices to demonstrate the concept of limiting reagent in acid–base reactions", *J. Chem. Educ.,* vol. 95, no. 2, pp. 305-309, 2018.
[http://dx.doi.org/10.1021/acs.jchemed.7b00410]

[16] W. Dungchai, O. Chailapakul, and C.S. Henry, "Use of multiple colorimetric indicators for paper-based microfluidic devices", *Anal. Chim. Acta,* vol. 674, no. 2, pp. 227-233, 2010.
[http://dx.doi.org/10.1016/j.aca.2010.06.019] [PMID: 20678634]

[17] D.A. Bruzewicz, M. Reches, and G.M. Whitesides, "Low-cost printing of poly(dimethylsiloxane) barriers to define microchannels in paper", *Anal. Chem.,* vol. 80, no. 9, pp. 3387-3392, 2008.
[http://dx.doi.org/10.1021/ac702605a] [PMID: 18333627]

[18] X. Li, J. Tian, G. Garnier, and W. Shen, "Fabrication of paper-based microfluidic sensors by printing", *Colloids Surf. B Biointerfaces,* vol. 76, no. 2, pp. 564-570, 2010.
[http://dx.doi.org/10.1016/j.colsurfb.2009.12.023] [PMID: 20097546]

[19] M.S. Khan, "Biosurface engineering through ink jet printing", *Colloids Surf B Biointerfaces.,* vol. 75, no. 2, pp. 441-447, 2010.
[http://dx.doi.org/10.1016/j.colsurfb.2009.09.032]

[20] G. Chitnis, Z. Ding, C-L. Chang, C.A. Savran, and B. Ziaie, "Laser-treated hydrophobic paper: an inexpensive microfluidic platform", *Lab on a Chip,* vol. 11, no. 6, pp. 1161-1165, 2011.
[http://dx.doi.org/10.1039/c0lc00512f]

[21] D. Gosselin, "Low-cost embossed-paper micro-channels for spontaneous capillary flow", In: *Sensors and Actuators B: Chemical* vol. 248. Elsevier. 395-401, 2017.
[http://dx.doi.org/10.1016/j.snb.2017.03.144]

[22] J. Yang, H. Li, T. Lan, L. Peng, R. Cui, and H.J.C.p. Yang, "Preparation, characterization, and properties of fluorine-free superhydrophobic paper based on layer-by-layer assembly", In: *Carbohydrate Polymers.* vol. 178. Elsevier. 228-237, 2017.
[http://dx.doi.org/10.1016/j.carbpol.2017.09.040]

[23] P. Zhang, L. Zhu, J. Cai, F. Lei, J.J. Qin, J. Xie, Y.M. Liu, Y.C. Zhao, X. Huang, L. Lin, M. Xia, M.M. Chen, X. Cheng, X. Zhang, D. Guo, Y. Peng, Y.X. Ji, J. Chen, Z.G. She, Y. Wang, Q. Xu, R. Tan, H. Wang, J. Lin, P. Luo, S. Fu, H. Cai, P. Ye, B. Xiao, W. Mao, L. Liu, Y. Yan, M. Liu, M. Chen, X.J. Zhang, X. Wang, R.M. Touyz, J. Xia, B.H. Zhang, X. Huang, Y. Yuan, R. Loomba, P.P.

Liu, and H. Li, "Association of iInpatient use of angiotensin-converting enzyme inhibitors and angiotensin II receptor blockers with mortality among patients with hypertension hospitalized with COVID-19", *Circ. Res.*, vol. 126, no. 12, pp. 1671-1681, 2020.
[http://dx.doi.org/10.1161/CIRCRESAHA.120.317134] [PMID: 32302265]

[24] M. Mohammadifar, J. Zhang, I. Yazgan, O.A. Sadik, and S.J.R.E. Choi, "Power-on-paper: Origami-inspired fabrication of 3-D microbial fuel cells", In: *Renewable Energy* vol. 118. Elsevier. 695-700, 2018.
[http://dx.doi.org/10.1016/j.renene.2017.11.059]

[25] M. Sher, R. Zhuang, U. Demirci, and W. Asghar, "Paper-based analytical devices for clinical diagnosis: recent advances in the fabrication techniques and sensing mechanisms", *Expert Rev. Mol. Diagn.*, vol. 17, no. 4, pp. 351-366, 2017.
[http://dx.doi.org/10.1080/14737159.2017.1285228] [PMID: 28103450]

[26] M. Wu, Q. Lai, Q. Ju, L. Li, H.D. Yu, and W. Huang, "Paper-based fluorogenic devices for *in vitro* diagnostics", *Biosens. Bioelectron.*, vol. 102, pp. 256-266, 2018.
[http://dx.doi.org/10.1016/j.bios.2017.11.006] [PMID: 29153947]

[27] J. F. d. S. Petruci, and A. A. Cardoso, "Portable and disposable paper-based fluorescent sensor for In Situ gaseous hydrogen sulfide determination in near real-time", *Analytical chemistry.*, vol. 88, no. 23, pp. 11714-11719, 2016.
[http://dx.doi.org/10.1021/acs.analchem.6b03325]

[28] X. Li, D.R. Ballerini, and W. Shen, "A perspective on paper-based microfluidics: Current status and future trends", *Biomicrofluidics,* vol. 6, no. 1, p. 011301, 2012.
[http://dx.doi.org/10.1063/1.3687398] [PMID: 22662067]

[29] M. Santhiago, C.C. Corrêa, J.S. Bernardes, M.P. Pereira, L.J.M. Oliveira, M. Strauss, and C.C.B. Bufon, "Flexible and foldable fully-printed carbon black conductive nanostructures on paper for high-performance electronic, electrochemical, and wearable devices", *ACS Appl. Mater. Interfaces,* vol. 9, no. 28, pp. 24365-24372, 2017.
[http://dx.doi.org/10.1021/acsami.7b06598] [PMID: 28650141]

[30] M. Punjiya, P. Mostafalu, S.R.J.I.B.C. Sonkusale, and S.C. Proceedings, "Low-cost paper-based electrochemical sensors with CMOS readout IC", 2014.
[http://dx.doi.org/10.1109/BioCAS.2014.6981728]

[31] J. Mettakoonpitak, "", "Electrochemistry on paper–based analytical devices", *RE:view,* vol. 28, pp. 1420-1436, 2016.

[32] P.J. Lamas-Ardisana, "Disposable electrochemical paper-based devices fully fabricated by screen-printing technique", In: *Electrochemistry Communications* vol. Vol. 75. Elsevier. 25-28, 2017.
[http://dx.doi.org/10.1016/j.elecom.2016.11.015]

[33] Z. Nie, "Electrochemical sensing in paper-based microfluidic devices", *Lab Chip.*, vol. 10, no. 4, pp. 477-483, 2010.
[http://dx.doi.org/10.1039/B917150A] [PMID: 20126688]

[34] L. Busa, S. Mohammadi, M. Maeki, A. Ishida, H. Tani, and M. Tokeshi, "Advances in microfluidic paper-based analytical devices for food and water analysis", *Micromachines (Basel),* vol. 7, no. 5, p. 86, 2016.
[http://dx.doi.org/10.3390/mi7050086] [PMID: 30404261]

[35] C. P. Kurup, N. F. Mohd-Naim, C. Tlili, and M. U. J. A. s. t. i. j. o. t. J. S. f. A. C. Ahmed, "A highly sensitive label-free aptasensor based on gold nanourchins and carbon nanohorns for the detection of lipocalin-2 (LCN-2)", *Analytical Sciences.*, vol. 37, no. 6, pp. 825-831, 2021.
[http://dx.doi.org/10.2116/analsci.20P303]

[36] W. Gao, Y. Liu, H. Zhang, and Z. Wang, "Electrochemiluminescence biosensor for nucleolin imaging in a single tumor cell combined with synergetic therapy of tumor", *ACS Sens.,* vol. 5, no. 4, pp. 1216-1222, 2020.

[http://dx.doi.org/10.1021/acssensors.0c00292] [PMID: 32223128]

[37]   J. Yu, S. Wang, L. Ge, and S. J. B. Ge, "A novel chemiluminescence paper microfluidic biosensor based on enzymatic reaction for uric acid determination", *Biosens Bioelectron.,* vol. 26, no. 7, pp. 3284-3289, 2011.
[http://dx.doi.org/10.1016/j.bios.2010.12.044] [PMID: 21257303]

[38]   L. Ge, S. Wang, X. Song, S. Ge, and J. Yu, "3D Origami-based multifunction-integrated immunodevice: low-cost and multiplexed sandwich chemiluminescence immunoassay on microfluidic paper-based analytical device", *Lab Chip,* vol. 12, no. 17, pp. 3150-3158, 2012.
[http://dx.doi.org/10.1039/c2lc40325k] [PMID: 22763468]

[39]   W.Y. Ko, T.J. Tien, C.Y. Hsu, and K.J. Lin, "Ultrasensitive label- and amplification-free photoelectric protocols based on sandwiched layer-by-layer plasmonic nanocomposite films for the detection of alpha-fetoprotein", *Biosens. Bioelectron.,* vol. 126, pp. 455-462, 2019.
[http://dx.doi.org/10.1016/j.bios.2018.11.020] [PMID: 30472442]

[40]   L. Cao, C. Huang, D. Cui Zhou, Y. Hu, T.M. Lih, S.R. Savage, K. Krug, D.J. Clark, M. Schnaubelt, L. Chen, F. da Veiga Leprevost, R.V. Eguez, W. Yang, J. Pan, B. Wen, Y. Dou, W. Jiang, Y. Liao, Z. Shi, N.V. Terekhanova, S. Cao, R.J.H. Lu, Y. Li, R. Liu, H. Zhu, P. Ronning, Y. Wu, M.A. Wyczalkowski, H. Easwaran, L. Danilova, A.S. Mer, S. Yoo, J.M. Wang, W. Liu, B. Haibe-Kains, M. Thiagarajan, S.D. Jewell, G. Hostetter, C.J. Newton, Q.K. Li, M.H. Roehrl, D. Fenyö, P. Wang, A.I. Nesvizhskii, D.R. Mani, G.S. Omenn, E.S. Boja, M. Mesri, A.I. Robles, H. Rodriguez, O.F. Bathe, D.W. Chan, R.H. Hruban, L. Ding, B. Zhang, H. Zhang, M. Amin, E. An, C. Ayad, T. Bauer, C. Birger, M.J. Birrer, S.M. Boca, W. Bocik, M. Borucki, S. Cai, S.A. Carr, S. Cerda, H. Chen, S. Chen, D. Chesla, A.M. Chinnaiyan, A. Colaprico, S. Cottingham, M. Derejska, S.M. Dhanasekaran, M.J. Domagalski, B.J. Druker, E. Duffy, M.A. Dyer, N.J. Edwards, M.J. Ellis, J. Eschbacher, A. Francis, J. Francis, S. Gabriel, N. Gabrovski, J. Gardner, G. Getz, M.A. Gillette, C.A. Goldthwaite Jr, P. Grady, S. Guo, P. Hariharan, T. Hiltke, B. Hindenach, K.A. Hoadley, J. Huang, C.D. Jones, K.A. Ketchum, C.R. Kinsinger, J.M. Koziak, K. Kusnierz, T. Liu, J. Long, D. Mallery, S. Mareedu, R. Matteotti, N. Maunganidze, P.B. McGarvey, P. Minoo, O.V. Paklina, A.G. Paulovich, S.H. Payne, O. Potapova, B. Pruetz, L. Qi, N. Roche, K.D. Rodland, D.C. Rohrer, E.E. Schadt, A.V. Shabunin, T. Shelton, Y. Shutack, S. Singh, M. Smith, R.D. Smith, L.J. Sokoll, J. Suh, R.R. Thangudu, S.X. Tsang, K.S. Um, D.R. Valley, N. Vatanian, W. Wang, G.D. Wilson, M. Wiznerowicz, Z. Zhang, and G. Zhao, "Proteogenomic characterization of pancreatic ductal adenocarcinoma", *Cell,* vol. 184, no. 19, pp. 5031-5052.e26, 2021.
[http://dx.doi.org/10.1016/j.cell.2021.08.023] [PMID: 34534465]

[41]   P. Jain, B. Chakma, S. Patra, and P. Goswami, "Potential biomarkers and their applications for rapid and reliable detection of malaria", *BioMed Res. Int.,* vol. 2014, pp. 1-20, 2014.
[http://dx.doi.org/10.1155/2014/852645] [PMID: 24804253]

[42]   R.G.A. Feachem, I. Chen, O. Akbari, A. Bertozzi-Villa, S. Bhatt, F. Binka, M.F. Boni, C. Buckee, J. Dieleman, A. Dondorp, A. Eapen, N. Sekhri Feachem, S. Filler, P. Gething, R. Gosling, A. Haakenstad, K. Harvard, A. Hatefi, D. Jamison, K.E. Jones, C. Karema, R.N. Kamwi, A. Lal, E. Larson, M. Lees, N.F. Lobo, A.E. Micah, B. Moonen, G. Newby, X. Ning, M. Pate, M. Quiñones, M. Roh, B. Rolfe, D. Shanks, B. Singh, K. Staley, J. Tulloch, J. Wegbreit, H.J. Woo, and W. Mpanju-Shumbusho, "Malaria eradication within a generation: ambitious, achievable, and necessary", *Lancet,* vol. 394, no. 10203, pp. 1056-1112, 2019.
[http://dx.doi.org/10.1016/S0140-6736(19)31139-0] [PMID: 31511196]

[43]   Y.W. Cheung, R.M. Dirkzwager, W.C. Wong, J. Cardoso, J. D'Arc Neves Costa, and J.A. Tanner, "Aptamer-mediated *Plasmodium*-specific diagnosis of malaria", *Biochimie,* vol. 145, pp. 131-136, 2018.
[http://dx.doi.org/10.1016/j.biochi.2017.10.017] [PMID: 29080831]

[44]   G. Dutta, "Electrochemical biosensors for rapid detection of malaria", In: *Materials Science for Energy Technologies* vol. 3. KeAi. 150-158, 2020.
[http://dx.doi.org/10.1016/j.mset.2019.10.003]

[45]   A. M. Ogunmolasuyi, R. Fogel, H. Hoppe, D. Goldring, and J. Limson, "A microfluidic paper analytical device using capture aptamers for the detection of PfLDH in blood matrices", *Malar J.,* vol. 21, no. 1, p. 174, 2022.
[http://dx.doi.org/10.1186/s12936-022-04187-6] [PMID: 35672848]

[46]   G. S. Alemayehu, "Genetic variation of *Plasmodium falciparum* histidine-rich protein 2 and 3 in Assosa zone, Ethiopia: its impact on the performance of malaria rapid diagnostic tests", *Malar J.,* vol. 20, no. 1, p. 394, 2021.
[http://dx.doi.org/10.1186/s12936-021-03928-3] [PMID: 34627242]

[47]   M.T. Makler, and D.J. Hinrichs, "Measurement of the lactate dehydrogenase activity of *Plasmodium falciparum* as an assessment of parasitemia", *Am. J. Trop. Med. Hyg.,* vol. 48, no. 2, pp. 205-210, 1993.
[http://dx.doi.org/10.4269/ajtmh.1993.48.205] [PMID: 8447524]

[48]   D. Tao, B. McGill, T. Hamerly, T. Kobayashi, P. Khare, A. Dziedzic, T. Leski, A. Holtz, B. Shull, A.E. Jedlicka, A. Walzer, P.D. Slowey, C.C. Slowey, S.E. Nsango, D.A. Stenger, M. Chaponda, M. Mulenga, K.H. Jacobsen, D.J. Sullivan, S.J. Ryan, R. Ansumana, W.J. Moss, I. Morlais, and R.R. Dinglasan, "A saliva-based rapid test to quantify the infectious subclinical malaria parasite reservoir", *Sci. Transl. Med.,* vol. 11, no. 473, p. eaan4479, 2019.
[http://dx.doi.org/10.1126/scitranslmed.aan4479] [PMID: 30602535]

[49]   T. E. Agbana, "Imaging & identification of malaria parasites using cellphone microscope with a ball lens", *PLoS One.,* vol. 13, no. 10, p. e0205020, 2018.
[http://dx.doi.org/10.1371/journal.pone.0205020] [PMID: 30286150]

[50]   J. J. Pollak, A. Houri-Yafin, and S. J. J. F. i. p. h. Salpeter, "Computer vision malaria diagnostic systems-progress and prospects", *Front Public Health.,* vol. 21, no. 5, p. 219, 2017.
[http://dx.doi.org/10.3389/fpubh.2017.00219] [PMID: 28879175]

CHAPTER 8

Enzymatic Biosensors and their Important Applications

Mridula Chaturvedi[1,*]

[1] *Amity Institute of Biotechnology, Amity University, Noida, Uttar Pradesh, India*

Abstract: Biosensors are an investigative contrivance used to find a chemical substance that combines an organic component with a physicochemical detector. Various biosensors are used in different fields: electrochemical-based, immune-based, magnetic-based, thermometric-based, acoustic-based, enzyme-based, optical-based, DNA-based, and tissue-based, *etc*. Enzyme-based biosensors are those which use enzymes or proteins for the recognition of elements and adjoining the inbuilt specificity of enzymes. Due to their high sensitivity and specificity, these enzymatic biosensors are frequently used in numerous fields, including the health and biomedical field. This book chapter will describe the various enzyme-based biosensors, their sensitivity and specificity, and their significant applications in different fields.

Keywords: Applications, Biosensors, Biomedical, Biotechnology, Enzymes, Quality indices.

INTRODUCTION

The arrangement of a biological element with a physiochemical element or appliance through a well-operational analytical device for the detection of a biological analysis is called a biosensor. A typical biosensor is principally comprised of an efficient transducer, a biological recognition element, and a signal processor, in which the biological recognition element can interact with the target molecule specifically, and then the transducer converts the biochemical information into a measurable signal [1].

Due to low instrumentation rate, high sensitivity, good accurateness, and ease to use, these biosensors are now gaining popularity in terms of commanding devices for sensing various analyses in different areas over conventional methods. As the physical, chemical, or biological reactions can be transformed into electrical,

---

[*] **Corresponding author Mridula Chaturvedi:** Amity Institute of Biotechnology, Amity University, Noida, Uttar Pradesh, India; E-mail: mridula.all@gmail.com

Vivek K. Chaturvedi, Dawesh P. Yadav and Mohan P. Singh (Eds.)

optical, or other signals by specific transducers, different kinds of biosensors are obtained according to the mode of signal transduction [2].

Similarly, the biosensors could also be classified based on the biological specificity conferring mechanism, the biocatalytic or affinity biosensors. In light of the monitoring analyses or reactions, the biosensors could be constituted by monitoring the concentration of analyses or reaction products directly or monitoring the inhibitor or activator of the bio-recognition element indirectly [3].

Thus on the discovery of these biosensors, a variety of bio-identified essentials like antibodies, nucleic acids, various fluids and different enzymes have been extensively used to make assorted biosensors. Historically, Clark and Lyons demonstrated the first biosensor *via* coupling the glucose. Determined by this revolutionary effort, the recognition of enzyme-based biosensors has full-grown extremely within the investigative population. As the key mechanism in the enzyme-based biosensors, enzymes possess outstanding bio-recognition potential and catalytic properties; that is, they could respond selectively to their matching substrates. Enzymes are not only the oldest but also the most common bio-recognition elements in current biosensors. In recent decades, enzyme-based biosensors have been proven to be a ground-breaking modus operandi in ailment finding, biological and biomedical study, and so forth [4].

Incorporated with dissimilar transducers, the enzyme-based biosensors could be primarily divided into electrochemical, optical, magnetic, tissue, DNA, RNA and other types. Given the speedy progress in this field, this chapter will introduce various enzyme-based biosensors and their important applications [5].

ENZYME-BASED BIOSENSORS

Enzymes are globular proteins largely composed of the 20 naturally occurring amino acids that can catalyze biochemical reactions and the recognized green catalysts with high specificity. The enzyme-based biosensor, as a precise type of biosensor, is also similarly made up of three parts; a bio-recognition element (enzyme), an effective transducer, and a digital signal processor. In other words, it depends on the collection of suitable enzymes as the bio-receptor molecules, skilled immobilization procedures, exact transducers, and lastly, integration of them in delicate forms to develop different kinds of biosensors. These well-established devices forever have high-quality immunity from meddling to complete the recognition of target biomolecules in a satisfactory range of concentrations [6].

The organization of an enzyme-based biosensor demands that the biocatalyst should be incorporated with the conductive electrodes so that the enzyme catalytic conversion in sequence can be transferred electronically. As a result, any electrical change at the conductive supporter, such as the depletion of reactants or the formation of products in the biocatalytic process, provides useful electronic transduction information of the biological reorganization event occurring at the electrode. On the other hand, the concentration of the target biomolecules is related to the decrement of the enzymatic product, as the enzymatic activity can be reserved by the detection objective [7]. It is of vital significance to immobilize the enzyme on the electrode during the manufacture of enzyme-based biosensors. The immobilization method must make sure the evenness of the active site, except to preserve the functionality and bioactivity of the biomolecules. Therefore, the immobilization technique can directly affect the sensitivity, selectivity, steadiness, and reproducibility of the constructed enzyme biosensor. With the requirement of a satisfactory enzyme biosensor, different kinds of immobilization techniques have been subjugated. These means can be classified into physical adsorption, covalent binding, entrapment within a polymerised matrix, and cross-linker [8]. Additionally, the biocatalytic activity of enzymes highly depends on the applied possible, temperature, and pH of the solution environment. All these factors can be credited to the concentration of substrate, the presence or absence of oxygen, and the properties of the enzyme. At the optimum pH and temperature, the greatest reproducibility and sensitivity are attained [9].

APPLICATIONS OF ENZYMATIC BIOSENSORS

Enzymatic biosensors can be used in different fields like the biomedical, food industry, biotechnology, microbiological, and healthcare field, *etc.*, as described below:

Enzyme Biosensors for Biomedical Applications

Amperometric Enzyme Biosensors

Amperometric enzyme biosensors are commonly divided into three main classes, or generations, depending on the electron transfer method used for the measurement of the biochemical reaction or the degree of separation of the biosensor components (transducer, enzyme, mediators and cofactors). In all cases, the presence of an enzyme is required, and therefore, sensor performance relies on different parameters, such as working pH and temperature [10].

First-Generation Biosensors

First-generation biosensors measure the concentration of analyses and/or products of enzymatic reactions that diffuse to the transducer surface and generate an electrical response. They are also called mediator-less Amperometric biosensors [11].

Table 1. List of some Oxidase enzymes and their substrates used inAmperometric biosensors.

S.No.	Enzyme	Source	Substrate	References
1.	Glucose oxidase	*Aspergillus niger*	β-D-Glucose	[12, 13]
2.	Glutamate oxidase	*Streptomyces sp*	L-Glutamate	[14]
3.	Alcohol oxidase	*Pichia polymorpha Hansenula pastoris*	Ethanol	[15]
4.	Lactate oxidase	*Pediococcus species Aerococcus viridians*	L-Lactate	[16]
5.	Ascorbate oxidase	*Cucurbita sp*	L-Ascorbic acid	[17]
6.	Cholesterol oxidase	*Streptomyces sp* porcine pancreas	Cholesterol	[18]
7.	Choline Oxidase	*Alcaligenes sp*	Choline Acetylcholine	
8.	Laccase	*Trametes pubescens Paraconiothyrium variable Trametes versicolor*	Polyphenols	[19, 20]
9.	*Mushroom*	Monophenols Bisphenol A	Dihydroxyphenols	[21]

Table 2. Few examples of Dehydrogenase enzymes and their sources used in Amperometric biosensors.

S.No.	Enzyme	Source	Substrate	References
1.	Alcohol dehydrogenase	*Saccharomyces cerevisiae* E.C. 1.1.1.1	Ethanol	[22, 23]
2.	Glutamate dehydrogenase	Bovine liver E.C. 1.4.1.2	L-Glutamate	[24]
3.	Glucose dehydrogenase	*Pseudomonas sp. Escherichia coli* EC 1.1.1.47	Glucose	[25]
4.	Lactate dehydrogenase	Rabbit muscle Chicken heart EC 1.1.1.27	L-Lactate	[26]

In oxidase enzymes, the most common co-factor is flavin adenine dinucleotide (FAD), which is not covalently bonded to the enzyme. Oxidase-based biosensors can either monitor the production of hydrogen peroxide (H_2O_2) by applying a fixed anodic potential (+0.7 V *vs.* Ag/AgCl) or oxygen (O_2) consumption by applying a fixed cathodic potential (−0.7 V *vs.* Ag/AgCl). An alternative approach, which is extensively used in the detection of hydrogen peroxide, is the introduction of peroxidases in the biosensor design, which allows for the detection of H_2O_2 by applying a low, reducing potential (Table **1**). Oxidase enzymes need molecular oxygen as a second substrate, so the oxidase-based biosensors are oxygen dependent. First-generation biosensors that use oxygen as an electron acceptor are thus subject to errors arising from changing or low concentrations of dissolved oxygen, impacting sensor response and reducing linearity. This oxygen dependence limits the applicability of the first generation of Amperometric biosensors in biological systems; for example, they are not suitable for use under ischemic conditions [27] (Table **2**).

Second-Generation Biosensors

Second-generation biosensors, also called mediator Amperometric biosensors, exploit mediators as oxidizing agents to act as electron carriers. This approach makes it possible to work at low potentials, avoiding O_2 dependence and the impact of interfering molecules. The most common and well-known mediators are ferricyanide and ferrocene, although methylene blue, phenazines, methyl violet, alizarin yellow, Prussian blue, thionin, azure A and C, toluidine blue and inorganic redox ions are also widely used. Further improvements are obtained by replacing oxygen with an electron acceptor capable of carrying electrons from the redox centre of the enzyme (E) to the electrode [28].

Third-Generation Biosensors

Third-generation biosensors rely on bioelectrocatalysis, where there is a direct electron transfer between enzyme and electrode. A third-generation biosensor consists of three elements: the enzyme as the bio-recognition element, the redox polymer (or the nano-scale wiring element) to ensure the signal propagation, and the electrode as the entrapping surface. Using a redox polymer to "wire" the redox centre of the sensing enzyme to the electrode surface improves performance. Third-generation biosensors are still being developed and are not commonly used for analysis. However, developments in polymer science and nanotechnology make third-generation biosensors promising, as the sensors are likely to have very short response times and be relatively independent of oxygen/cofactor concentrations [29].

Enzyme Biosensors in Biological Matrices: Composition and Matrix-Related Detection Problems

In biosensing, when the detecting device is directly in contact with biological fluids, undesired signals from interferons are a serious and prevailing problem. There are several low and high-molecular-weight interfering compounds, such as oxidizable acids (*e.g.*, ascorbic acid, uric acid, homovanillic acid) and bases (*e.g.*, oxidizable catecholamines and indoleamine). Biological fluids may also contain drugs and their metabolites (*e.g.*, acetaminophen). Moreover, proteins (high molecular weight interferents) at high concentrations can absorb nonspecifically to the transducer surface, interfering with the detection of target molecules, which are often present at low concentrations [30].

All biological matrices can contain compounds that cause biofouling or enzyme inactivation: principally, these kinds of molecules are low and high molecular-weight proteins, but may also include small water-soluble molecules (*e.g.*, sugars) and hydrophobic compounds (*e.g.*, lipids). All of these compounds can cause electrode passivation and/or biofouling. To face the issue of sensor impairment, several membrane coatings, such as Nafion, polyurethanes with phospholipid polar groups, 2-methacryloyloxyethyl phosphorylcholine, hyaluronic acid, humic acids, phosphorylcholine and polyvinyl alcohol hydrogels, have been used. These membranes reduce the impact of biological fluids on biosensors, minimizing protein adsorption while still allowing target analytes to reach the sensing surface. However, such strategies are less effective against low molecular weight protein fragments and large charged cell deposits [31].

Table 3. Qualitative composition of selected biological fluids.

Fluid	Cations	Anions	Proteins	Metabolites	Nutrients
Saliva	++	+++	++	---	---
Urine	++	+++	--	+++	----
Blood	++	++	+++	+++	+++
ECF	++++	+++++	--	++	+
Tears	++	++	--	+	+
Sweat	+++	+++	--	+	+

Mono- and divalent cations are present at high concentrations in biological fluids and can activate or inhibit many enzymes (Table **3**). These cations can act as allosteric effectors without participating in the enzymatic reaction or altering the conformation needed for catalytic activity. The charge and the size of the ion present in the catalytic site of an enzyme represent one of the important issues that determine which metal ion can inhibit an enzyme. In some cases, the interaction of monovalent cations with enzymes is important for catalysis; K^+ and Na^+, for example, stabilize glucose oxidase against thermal denaturation. Electrode passivation can arise with the non-specific adsorption of proteins and lipids. Deposition of polymeric films, such as *o*-phenylenediamine, polyeugenol, polypyrrole and other films (conducting or nonconducting) or specific membranes, such as Nafion, polymethyl cellulose, or hydrogels, can prevent passivation, although often at the expense of response time. Carbon nanotubes have also been used to minimize electrode fouling, for example, from oxidation of NADH [32].

Saliva

Saliva is a clear, slightly acidic exocrine secretion that contains oral bacteria and food debris. It is composed of several compounds, including cations, such as sodium, potassium, calcium, magnesium, and anions, such as bicarbonate, and phosphates. Saliva also contains nitrogenous compounds, such as urea and ammonia, as well as immunoglobulin, proteins, enzymes and mucins. Saliva glucose concentrations range approximately from 20 to 200 µmol/L in normal and diabetic patients; they closely follow circadian blood glucose fluctuations. Normal saliva is a complex solution derived from parotid, submandibular, sublingual and minor gland secretions. It may also include bacteria, leukocytes, epithelial cells, and gingival crevicular fluid, which make the measurement of components of saliva difficult. Concentrations of alcohol, glucose and lactate measured in saliva correlate well with concentrations in blood serum. Saliva also contains enzymes, such as amylase, which transforms complex sugars into simple ones; thus, its activity could interfere with the accurate detection of some analytes, such as glucose. Normal saliva pH ranges from 6.5 to 7.4 and is related to saliva buffers (bicarbonate) and oral hygiene [33].

Urine

Na^+, K^+, Ca^{2+}, Mg^{2+} and NH_4^+ account for nearly all the cations present in urine, while chloride, sulphate, phosphate and bicarbonate account for about 80% of the anions present. Cells and proteins are negligible. Urine pH normally ranges between 4.6 and 8 and is related to diet and overall health.

Blood, Plasma and Serum

Blood is a complex mixture of plasma (the liquid component), white blood cells, red blood cells, and platelets. Plasma is composed of 90% water and represents about 55% of blood volume. The serum is blood plasma without fibrinogen. In serum and plasma, several water-soluble compounds are present, such as nutrients, hormones and electrolytes. It is also possible to find drugs and proteins, such as globulins (including antibodies), fibrinogen (blood clotting factor), albumin (major protein constituent) and other clotting factors. There may be up to 10,000 proteins in serum, including immunoglobulins, albumin, lipoproteins, haptoglobin, and transferrin. Blood is a highly buffered fluid, with normal blood pH ranging from 7.35 to 7.45 [34].

Extracellular Fluid (ECF) and Brain Extracellular Fluid (bECF)

ECF is mainly composed of ions (Na^+, Cl^- and Ca^{2+}), glucose, amino acids and ATP, with negligible protein content. One of the major applications of amperometric biosensors involves their implantation in specific brain regions, in direct contact with brain extracellular fluid (bECF). *In vivo* microdialysis allowed a deep characterisation of bECF composition. In particular, the ionic composition of bECF is well known (NaCl 147 mM, KCl 2.7 mM, $CaCl_2$ 1.2 mM, $MgCl_2$ 0.85 mM). Electroactive molecules are also present, such as ascorbic acid (AA). AA performs various functions at the neuronal level and is considered to be one of the most powerful antioxidants. Also, catecholamines, such as dopamine, noradrenaline and their major metabolites, such as 3,4-dihydroxyphenylacetic acid, 3-methoxytyramine and homovanillic acid (HVA), are present. In bECF, uric acid produced by the catabolism of purines may be detected, as well as 5-hydroxy-tryptamine (serotonin) and its metabolite 5-hydroxyindoleacetic. bECF also contains many other molecules which can be detected by employing amperometric biosensors, such as glucose, lactate, glutamate, acetylcholine and choline [35].

Tears

Tears are produced by the lachrymal glands and can be used as an interesting fluid for non-invasive monitoring. The amount of protein present in tears is negligible. The normal pH range is between 6.5 and 7.6.

Sweat

Sweat includes urea, uric acid, sugar, lactic acid, amino acids and ammonia; concentrations vary widely from person to person.

Changes in Biological Fluids Composition Related to Physiological and Pathological Conditions

The composition of the biological fluids described above may change in response to various physiological and pathological conditions. The pH of saliva and urine, for example, is closely related to diet; saliva pH can drop immediately after food consumption, and low pH is related to poor oral hygiene. Some pathological conditions, such as inflammation, neurodegenerative diseases, infections, or cancer, can also modify fluid parameters, such as dissolved oxygen percentage, chemical composition and pH. These variations in the composition of biological fluids can impact biosensor performance. For example, ischemia would decrease the response of first-generation biosensors under their oxygen dependence, and over-production of Reactive Oxygen Species (ROS) and proteases could damage biosensors. Such variations in biological fluid composition are known. For example, a PET study confirmed ROS increase with striatal oxidative stress in patients affected by Parkinson's disease. Similarly, cancer, inflammation and infectious diseases result in increased ROS production. Notably, ROS are primarily responsible for biosensor ageing and enzyme-related loss of sensitivity due to the inactivation of catalytic sites or cofactors (reduction of apparent V_{MAX}) or the oxidation of non-catalytic sites with consequent molecular rearrangement and reduced substrate affinity (an increase of apparent K_M). Neurodegeneration, inflammation and cancer are also responsible for the increase in proteases that may damage biosensors [36].

Application of Enzyme Biosensors in Analysis of Food and Beverages

Analysing different parameters in food products and monitoring a production process requires quick and reliable analytical methods and devices. For this purpose, biosensors can be a suitable option, whereas most of the current quality control techniques are time-consuming, expensive, and unpractical. In this paper, we describe biosensors developed for the analysis of different components present in food samples, namely, glucose, fructose, sucrose, lactose, lactic, malic, acetic, ascorbic, and citric and amino acids, ethanol, glycerol, and triglyceride. Biosensors showed desirable sensitivity, selectivity, and response time required for various applications. They are often designed to avoid interference from components present in a complex sample to be analyzed. Quality control during the manufacturing process and testing of qualitative parameters of food materials and final food products is a very important task for manufacturers and hygiene inspections [37]. Duration and accuracy of analysis play a key role in this case. One option how to perform an analysis of microbial contamination or important production parameters is to use biosensor devices. The use of biosensors in foods can be broadly divided into two groups: enzyme sensors for food components, and immunosensors for pathogenic microbes and pesticides. They are used in the food industry to obtain proximate analysis, nutritional labelling, determination of pesticide residues, naturally occurring toxins and anti-nutrients, processing changes, microbial contamination, enzymatic inactivation, and biochemical oxygen demand of wastes. The food industry and biotechnology are the fields where biosensor applications recently penetrated, though not as intensively as in the field of medical diagnostics. One of the reasons can be that while in the medical area, the main matrices are blood, serum, or urine, in the food industry sector, there are more types of samples and variations in their compositions. This makes the process of biosensor design, unification, and optimization of measurement conditions more difficult. A biosensor is an analytical device which converts a biological response into an electrical signal. It consists of two main components: a bioreceptor or a biorecognition element, which recognizes the target analyte and a transducer, which converts the recognition event into a measurable electrical signal. As a biorecognition system, enzymes, antibodies, DNA, microorganism, *etc.*, can be used. A majority of biosensors existing today use three types of transducers for converting the action of the bioreceptor molecule into a measurable signal [38]. These are mainly amperometry based on H_2O_2 or O_2 measurement, potentiometry based on pH or pIon measurement, photometry utilizing optical fibres, and calorimetric biosensors measuring temperature change. The most commonly used class of biosensors are electrochemical-based ones.

Enzyme biosensors for point-of-care testing

Biosensors are devices that integrate a variety of technologies, containing biology, electronics, chemistry, physics, medicine, informatics, and correlated technology. Biosensors act as a transducer with a biorecognition element and transform a biochemical reaction on the transducer surface directly into a measurable signal. The biosensors have the advantages of rapid analysis, low cost, and high precision, which are widely used in many fields, such as medical care, disease diagnosis, food detection, environmental monitoring, and the fermentation industry. The enzyme biosensors show excellent application value owing to the development of fixed technology and the characteristics of specific identification, which can be combined with point-of-care testing technology (POCT). POCT technology is attracting more and more attention as a very effective method of clinic detection [39].

Enzymatic Biosensors in Biotechnology

A biosensor is a device that has the potential to detect a particular substance or analyte with high specificity. Examples of such analytes include glucose, lactate, glutamate and glutamine. Most biosensors are capable of measuring the concentration of an analyte in an aqueous solution, generally producing an electrical signal, which is considered to be proportional to the analyte concentration in its measuring range. An enzymatic biosensor comprises an enzyme, which recognizes and then reacts with the target analytes producing a chemical signal, a transducer, which produces a physical signal out of that chemical one, and an electronic amplifier, which conditions and then amplifies the signal [40].

Biosensors permit the analysis of complex biological media. The detection of a huge number of compounds is of immense relevance for scientific research and also for process control in the food and chemical industry. It is also indispensable in the health care field for the diagnosis and subsequent treatment of diseases and the monitoring of illnesses. The biotechnology and pharmaceutical industries greatly desire frequent, continuous analysis of biological media.

Such analyses are performed with the help of analytical instruments such as HPLC systems, which, although reliable and robust, are expensive and have limited suitability for online operation. For this reason, the acquisition of Jobst Technologies GmbH places IST AG as a key provider of reliable and high-performance online biosensors [41].

Enzymatic Biosensors for Metabolic Parameters

Back in the early 1960s, Clark and Lyons launched the first glucose sensor employing an enzyme (glucose oxidase, GOx) as the receptor, this enzyme being specific for glucose. Enzymes allow the highly specific measurement of their corresponding analyte even in complex mixtures like blood and fermentation broth; it is like finding a needle in a haystack. Analytes such as lactate, glucose, glutamate, and glutamine play a vital role in the metabolism of living organisms. Glutamine and glucose support cell function and growth; lactate is developed by cells and allows judging the efficiency of the cells. Such analyses are performed with the help of analytical instruments such as HPLC systems, which, although reliable and robust, are expensive and have limited suitability for online operation. For this reason, the acquisition of Jobst Technologies GmbH places IST AG as a key provider of reliable and high-performance online biosensors [42].

Use of Enzymatic Biosensors as Quality Indices

In recent decades increased knowledge about the biological capacity of enzymes has made it possible to create a new generation of products and processes. Among these products are notably biosensors, which represent a powerful alternative to conventional analytical techniques. This technology has advanced considerably in recent years, basically because of the creation of devices applied in the area of biomedicine. These advanced technologies have been gradually transferred horizontally to other sectors, such as the environment and the agro-food industry. A biosensor is defined as a compact device for analysis that incorporates a biological or biomimetic recognition element (nucleic acid, enzyme, anti-body, receptor, tissue, and cell) associated with a transduction system that allows for processing the signal produced by the interaction between the recognition element and the analytes. The principle of detection of a biosensor is based on the specific interaction between the analytes of interest and the recognition element [43]. As a result of this specific interaction, changes are produced in one or several physical-chemical properties (pH, electron transference, heat transference, change of potential or mass, variation of optical properties, *etc.*). These changes are detected and can be measured by a transductor. This system transforms the response of the recognition element into an electronic signal indicative of the presence of the analyte under study or proportional to its concentration in the sample. Biosensors currently represent powerful tools for analysis with numerous applications in the agro-food industry, mainly in biotechnological instruments. The most important characteristics of these devices to be competitive with other technologies in the agro-food industry are their specificity, high sensitivity, short response time, their capacity to be incorporated into integrated systems, the facility to automate them, their capacity to work in real-time, their versatility and low production cost [44]. In recent years, the number of scientific investigations and reviews on biosensors has been very high, which reflects the considerable interest in the theme. Ironically, there is a lag between the high level of scientific and technological development and the limited use of these devices in the agro-food sector basically because of structural characteristics of the sector, such as legislation, methodological inertia, absorption capacity and environmental factors [45].

CONCLUSION

Amperometric enzyme-based biosensors are complex analytical devices combining interdisciplinary knowledge, based on electrochemistry, materials science, polymer synthesis, enzymology and biological chemistry. Although biosensors can be fully characterized in a controlled laboratory environment, use in biological matrices, such as biological fluids, is more challenging. This latter step shows the research on biosensors (development) and the use of biosensors for research as diagnosis and follow-up analytical devices.

LIST OF ABBREVIATIONS

FAD	Flavin adenine dinucleotide
H_2O_2	Hydrogen peroxide
Ag/AgCl	Silver chloride
K^+/Na^+	Potassium/Sodium
NADH	Nicotinamide adenine dinucleotide
Ca^{2+}	Calcium
Mg^{2+}	Magnesium
NH_4^+	Ammonium
bECF	Brain extracellular fluid
NaCl	Sodium chloride
KCl	Potassium chloride
$CaCl_2$	Calcium chloride
$MgCl_2$	Magnesium chloride
AA	Ascorbic acid
HVA	Homovanillic acid
ROS	Reactive oxygen species
POCT	Point-of-care testing technology

HPLC High-performance liquid chromatography

GOx Glucose oxidase

ACKNOWLEDGEMENTS

This work was supported by a grant provided by the Indian Council of Medical Research, Govt. of India under a project file no (BMI/11(50)/2022) with proposal id 2021-11340-F1. The author is grateful for the suggestions given by Dr. Abhishek Chaturvedi.

REFERENCES

[1] A Hasan, M Nurunnabi, M Morshed, A Paul, A Polini, T Kuila, M Al Hariri, Y.K Lee, and A.A. Jaffa, "Recent advances in application of biosensors in tissue engineering", *Biomed Res Int.,* vol. 2014, p. 307519, 2014.
[http://dx.doi.org/10.1155/2014/307519] [PMID: 25165697]

[2] D. Grieshaber, R. MacKenzie, J. Vörös, and E. Reimhult, "Electrochemical biosensors - sensor principles and architectures", *Sensors (Basel),* vol. 8, no. 3, pp. 1400-1458, 2008.
[http://dx.doi.org/10.3390/s80314000] [PMID: 27879772]

[3] R.S. Lowe, "Overview of biosensor and bioarray technologies", In: *Handbook of Bios. and Biochips.* Wiley: Weinheim, Germany, 2007.
[http://dx.doi.org/10.1002/9780470061565.hbb003]

[4] L. Clark, "Electrode systems for continuous monitoring in cardiovascular surgery", *Ann. N.Y. Acad. Sci.,* vol. 102, pp. 29-45, 1962.
[http://dx.doi.org/10.1111/j.1749-6632.1962.tb13623.x.196]

[5] M. Pohanka, and P. Skládal, "Electrochemical biosensors - principles and applications", *J. Appl. Biomed.,* vol. 6, no. 2, pp. 57-64, 2008.
[http://dx.doi.org/10.32725/jab.2008.008]

[6] P. D'Orazio, "Biosensors in clinical chemistry", *Clin. Chim. Acta,* vol. 334, no. 1-2, pp. 41-69, 2003.
[http://dx.doi.org/10.1016/S0009-8981(03)00241-9] [PMID: 12867275]

[7] M.S. Wilson, "Electrochemical immunosensors for the simultaneous detection of two tumor markers", *Anal. Chem.,* vol. 77, no. 5, pp. 1496-1502, 2005.
[http://dx.doi.org/10.1021/ac0485278] [PMID: 15732936]

[8] J. Wang, "Real-time electrochemical monitoring: toward green analytical chemistry", *Acc. Chem. Res.,* vol. 35, no. 9, pp. 811-816, 2002.
[http://dx.doi.org/10.1021/ar010066e] [PMID: 12234211]

[9] J. Švorc, S. Miertuš, J. Katrlík, and M. Stred'anský, "Composite transducers for amperometric biosensors. The glucose sensor", *Anal. Chem.,* vol. 69, no. 11, pp. 2086-2090, 1997.
[http://dx.doi.org/10.1021/ac9609485] [PMID: 21639250]

[10] Z. Zhu, L. Garcia-Gancedo, A.J. Flewitt, H. Xie, F. Moussy, and W.I. Milne, "A critical review of glucose biosensors based on carbon nanomaterials: carbon nanotubes and graphene", *Sensors (Basel),* vol. 12, no. 5, pp. 5996-6022, 2012.
[http://dx.doi.org/10.3390/s120505996] [PMID: 22778628]

[11] J.M. Pingarrón, P. Yáñez-Sedeño, and A. González-Cortés, "Gold nanoparticle-based electrochemical biosensors", *Electrochim. Acta,* vol. 53, no. 19, pp. 5848-5866, 2008.
[http://dx.doi.org/10.1016/j.electacta.2008.03.005]

[12] R.D. O'Neill, S.C. Chang, J.P. Lowry, and C.J. McNeil, "Comparisons of platinum, gold, palladium

and glassy carbon as electrode materials in the design of biosensors for glutamate", *Biosens. Bioelectron.,* vol. 19, no. 11, pp. 1521-1528, 2004.
[http://dx.doi.org/10.1016/j.bios.2003.12.004] [PMID: 15093225]

[13]   W. Zhou, P-J. Jimmy Huang, J. Ding, and J. Liu, "Aptamer-based biosensors for biomedical diagnostics", *Analyst (Lond.),* vol. 139, no. 11, pp. 2627-2640, 2014.
[http://dx.doi.org/10.1039/c4an00132j] [PMID: 24733714]

[14]   J Castillo, S Gáspár, S Leth, M Niculescu, A Mortari, I Bontidean, V Soukharev, S.A Dorneanu, and A. Ryabov, "Biosensors for life quality: Design, development and applications", *Sens. Actuators B Chem.,* vol. 102, no. 2, pp. 179-194, 2004.
[http://dx.doi.org/10.1016/j.snb.2004.04.084]

[15]   M. Belluzo, M. Ribone, and C. Lagier, "Assembling amperometric biosensors for clinical diagnostics", *Sensors (Basel),* vol. 8, no. 3, pp. 1366-1399, 2008.
[http://dx.doi.org/10.3390/s8031366] [PMID: 27879771]

[16]   S.R. Corrie, J.W. Coffey, J. Islam, K.A. Markey, and M.A.F. Kendall, "Blood, sweat, and tears: developing clinically relevant protein biosensors for integrated body fluid analysis", *Analyst (Lond.),* vol. 140, no. 13, pp. 4350-4364, 2015.
[http://dx.doi.org/10.1039/C5AN00464K] [PMID: 25909342]

[17]   Y.C. Wang, A.L. Stevens, and J. Han, "Million-fold preconcentration of proteins and peptides by nanofluidic filter", *Anal. Chem.,* vol. 77, no. 14, pp. 4293-4299, 2005.
[http://dx.doi.org/10.1021/ac050321z] [PMID: 16013838]

[18]   R. Ramasamy, N. Gopal, V. Kuzhandaivelu, and S.B. Murugaiyan, "Biosensors in clinical chemistry: An overview", *Adv. Biomed. Res.,* vol. 3, no. 1, p. 67, 2014.
[http://dx.doi.org/10.4103/2277-9175.125848] [PMID: 24627875]

[19]   Y.H. Lee, and R. Mutharasan, "Sensors technology handbook", In: *Elsevier* Biosensors: Amsterdam, Netherlands, pp. 161-180, 2005..
[http://dx.doi.org/10.1016/B978-075067729-5/50046-X]

[20]   S.V. Dzyadevych, V.N. Arkhypova, A.P. Soldatkin, A.V. El'skaya, C. Martelet, and N. Jaffrezic-Renault, "Amperometric enzyme biosensors: Past, present and future", *IRBM,* vol. 29, no. 2-3, pp. 171-180, 2008.
[http://dx.doi.org/10.1016/j.rbmret.2007.11.007]

[21]   A. Harper, and M.R. Anderson, "Electrochemical glucose sensors--developments using electrostatic assembly and carbon nanotubes for biosensor construction", *Sensors (Basel),* vol. 10, no. 9, pp. 8248-8274, 2010.
[http://dx.doi.org/10.3390/s100908248] [PMID: 22163652]

[22]   S. Palanisamy, B. Unnikrishnan, and S. Chen, "An amperometric biosensor based on direct immobilization of horseradish peroxidase on electrochemically reduced graphene oxide modified screen printed carbon electrode", *Int. J. Electrochem. Sci.,* vol. 7, pp. 7935-7947, 2012.

[23]   J. Wang, "Electrochemical glucose biosensors", *Chem. Rev.,* vol. 108, no. 2, pp. 814-825, 2008.
[http://dx.doi.org/10.1021/cr068123a] [PMID: 18154363]

[24]   C.P. McMahon, G. Rocchitta, P.A. Serra, S.M. Kirwan, J.P. Lowry, and R.D. O'Neill, "Control of the oxygen dependence of an implantable polymer/enzyme composite biosensor for glutamate", *Anal. Chem.,* vol. 78, no. 7, pp. 2352-2359, 2006.
[http://dx.doi.org/10.1021/ac0518194] [PMID: 16579619]

[25]   M.F. Hossain, and J.Y. Park, "Plain to point network reduced graphene oxide - activated carbon composites decorated with platinum nanoparticles for urine glucose detection", *Sci. Rep.,* vol. 6, no. 1, p. 21009, 2016.
[http://dx.doi.org/10.1038/srep21009] [PMID: 26876368]

[26]   Ö. Sağlam, B. Kızılkaya, H. Uysal, and Y. Dilgin, "Biosensing of glucose in flow injection analysis

system based on glucose oxidase-quantum dot modified pencil graphite electrode", *Talanta,* vol. 147, pp. 315-321, 2016.
[http://dx.doi.org/10.1016/j.talanta.2015.09.050] [PMID: 26592613]

[27]   R. Devasenathipathy, V. Mani, S.M. Chen, S.T. Huang, T.T. Huang, C.M. Lin, K.Y. Hwa, T.Y. Chen, and B.J. Chen, "Glucose biosensor based on glucose oxidase immobilized at gold nanoparticles decorated graphene-carbon nanotubes", *Enzyme Microb. Technol.,* vol. 78, pp. 40-45, 2015.
[http://dx.doi.org/10.1016/j.enzmictec.2015.06.006] [PMID: 26215343]

[28]   G. Rocchitta, O. Secchi, M.D. Alvau, D. Farina, G. Bazzu, G. Calia, R. Migheli, M.S. Desole, R.D. O'Neill, and P.A. Serra, "Simultaneous telemetric monitoring of brain glucose and lactate and motion in freely moving rats", *Anal. Chem.,* vol. 85, no. 21, pp. 10282-10288, 2013.
[http://dx.doi.org/10.1021/ac402071w] [PMID: 24102201]

[29]   Ş. Şimşek, E. Aynacı, and F. Arslan, "An amperometric biosensor for L-glutamate determination prepared from L-glutamate oxidase immobilized in polypyrrole-polyvinylsulphonate film", *Artif. Cells Nanomed. Biotechnol.,* vol. 44, no. 1, pp. 356-361, 2016.
[http://dx.doi.org/10.3109/21691401.2014.951723] [PMID: 25682838]

[30]   O. Soldatkin, A. Nazarova, N. Krisanova, A. Borysov, D. Kucherenko, I. Kucherenko, N. Pozdnyakova, A. Soldatkin, and T. Borisova, "Monitoring of the velocity of high-affinity glutamate uptake by isolated brain nerve terminals using amperometric glutamate biosensor", *Talanta,* vol. 135, pp. 67-74, 2015.
[http://dx.doi.org/10.1016/j.talanta.2014.12.031] [PMID: 25640127]

[31]   C.P. McMahon, G. Rocchitta, P.A. Serra, S.M. Kirwan, J.P. Lowry, and R.D. O'Neill, "The efficiency of immobilised glutamate oxidase decreases with surface enzyme loading: an electrostatic effect, and reversal by a polycation significantly enhances biosensor sensitivity", *Analyst (Lond.),* vol. 131, no. 1, pp. 68-72, 2006.
[http://dx.doi.org/10.1039/B511643K] [PMID: 16365665]

[32]   S.R. Chinnadayyala, M. Santhosh, N.K. Singh, and P. Goswami, "Alcohol oxidase protein mediated *in-situ* synthesized and stabilized gold nanoparticles for developing amperometric alcohol biosensor", *Biosens. Bioelectron.,* vol. 69, pp. 155-161, 2015.
[http://dx.doi.org/10.1016/j.bios.2015.02.015] [PMID: 25725464]

[33]   M. Gamella, S. Campuzano, J. Manso, G.G. Rivera, F. López-Colino, A.J. Reviejo, and J.M. Pingarrón, "A novel non-invasive electrochemical biosensing device for *in situ* determination of the alcohol content in blood by monitoring ethanol in sweat", *Anal. Chim. Acta,* vol. 806, pp. 1-7, 2014.
[http://dx.doi.org/10.1016/j.aca.2013.09.020] [PMID: 24331037]

[34]   O. Secchi, M. Zinellu, Y. Spissu, M. Pirisinu, G. Bazzu, R. Migheli, M. Desole, R. O'Neill, P. Serra, and G. Rocchitta, "Further *in-vitro* characterization of an implantable biosensor for ethanol monitoring in the brain", *Sensors (Basel),* vol. 13, no. 7, pp. 9522-9535, 2013.
[http://dx.doi.org/10.3390/s130709522] [PMID: 23881145]

[35]   G. Rocchitta, O. Secchi, M.D. Alvau, R. Migheli, G. Calia, G. Bazzu, D. Farina, M.S. Desole, R.D. O'Neill, and P.A. Serra, "Development and characterization of an implantable biosensor for telemetric monitoring of ethanol in the brain of freely moving rats", *Anal. Chem.,* vol. 84, no. 16, pp. 7072-7079, 2012.
[http://dx.doi.org/10.1021/ac301253h] [PMID: 22823474]

[36]   P. Giménez-Gómez, M. Gutiérrez-Capitán, F. Capdevila, A. Puig-Pujol, C. Fernández-Sánchez, and C. Jiménez-Jorquera, "Monitoring of malolactic fermentation in wine using an electrochemical bienzymatic biosensor for l-lactate with long term stability", *Anal. Chim. Acta,* vol. 905, pp. 126-133, 2016.
[http://dx.doi.org/10.1016/j.aca.2015.11.032] [PMID: 26755146]

[37]   N. Hernández-Ibáñez, L. García-Cruz, V. Montiel, C.W. Foster, C.E. Banks, and J. Iniesta, "Electrochemical lactate biosensor based upon chitosan/carbon nanotubes modified screen-printed graphite electrodes for the determination of lactate in embryonic cell cultures", *Biosens. Bioelectron.,*

vol. 77, pp. 1168-1174, 2016.
[http://dx.doi.org/10.1016/j.bios.2015.11.005] [PMID: 26579934]

[38]   L. Andrus, R. Unruh, N. Wisniewski, and M. McShane, "Characterization of lactate sensors based on lactate oxidase and palladium benzoporphyrin immobilized in hydrogels", *Biosensors (Basel),* vol. 5, no. 3, pp. 398-416, 2015.
[http://dx.doi.org/10.3390/bios5030398] [PMID: 26198251]

[39]   Y. Wen, J. Xu, M. Liu, D. Li, and H. He, "Amperometric vitamin C biosensor based on the immobilization of ascorbate oxidase into the biocompatible sandwich-type composite film", *Appl. Biochem. Biotechnol.,* vol. 167, no. 7, pp. 2023-2038, 2012.
[http://dx.doi.org/10.1007/s12010-012-9711-y] [PMID: 22644641]

[40]   M. Liu, Y. Wen, J. Xu, H. He, D. Li, R. Yue, and G. Liu, "An amperometric biosensor based on ascorbate oxidase immobilized in poly(3,4-ethylenedioxythiophene)/multi-walled carbon nanotubes composite films for the determination of L-ascorbic acid", *Anal. Sci.,* vol. 27, no. 5, pp. 477-482, 2011.
[http://dx.doi.org/10.2116/analsci.27.477] [PMID: 21558652]

[41]   K. Lata, V. Dhull, and V. Hooda, "Fabrication and optimization of ChE/ChO/HRP-AuNPs/-MWCNTs based silver electrode for determining total cholesterol in serum", *Biochem. Res. Int.,* vol. 2016, pp. 1-11, 2016.
[http://dx.doi.org/10.1155/2016/1545206] [PMID: 26885393]

[42]   V. Aggarwal, J. Malik, A. Prashant, P.K. Jaiwal, and C.S. Pundir, "Amperometric determination of serum total cholesterol with nanoparticles of cholesterol esterase and cholesterol oxidase", *Anal. Biochem.,* vol. 500, pp. 6-11, 2016.
[http://dx.doi.org/10.1016/j.ab.2016.01.019] [PMID: 26853742]

[43]   S.K. Shukla, A.P.F. Turner, and A. Tiwari, "Cholesterol oxidase functionalised polyaniline/carbon nanotube hybrids for an amperometric biosensor", *J. Nanosci. Nanotechnol.,* vol. 15, no. 5, pp. 3373-3377, 2015.
[http://dx.doi.org/10.1166/jnn.2015.10209] [PMID: 26504954]

[44]   M.M. Rahman, and A.M. Asiri, "Selective choline biosensors based on choline oxidase co-immobilized into self-assembled monolayers on micro-chips at low potential", *Anal. Methods,* vol. 7, no. 22, pp. 9426-9434, 2015.
[http://dx.doi.org/10.1039/C5AY02456K]

[45]   A.T. Tunç, E. Aynacı Koyuncu, and F. Arslan, "Development of an acetylcholinesterase-choline oxidase based biosensor for acetylcholine determination", *Artif. Cells Nanomed. Biotechnol.,* pp. 1-6, 2015.
[PMID: 26367252] [http://dx.doi.org/10.3109/21691401.2015.1080167]

CHAPTER 9

Recent Trends of Nanobiosensor in Agriculture

Dhitri Borah[1], Ravi Kumar Goswami[2], Avanish Kumar Shrivastav[3], Priti Giri[4], Neelesh Kumar[3] and Tejveer Singh[5,*]

[1] *Department of Zoology, Biswanath College, Biswanath Chariali, 784176, Assam, India*

[2] *Department of Zoology, Hindu College, University of Delhi, New Delhi- 110007 India*

[3] *Department of Biotechnology, Delhi Technological University, Delhi, 110042, India*

[4] *Department of Botany, University of Delhi, Delhi-110007, India*

[5] *Translational Oncology Laboratory, Department of Zoology, Hansraj College, University of Delhi, New Delhi, Delhi-110067, India*

Abstract: Sustainable agriculture has the potential to benefit greatly from nanobiosensors. Nanomaterials are crucial components of numerous biotic and abiotic remediation systems and have a huge impact on the mobility, fate and toxicity of soil contaminants in agriculture. It is believed that nanobiosensor have a revolutionary impact on the field of agriculture by focusing research and development toward the goals of achieving sustainable agriculture. Nanobiosensors have significant benefits such as improved recognition sensitivity or specificity, and retain immense promise for the application of nanobiosensor in various areas such as food quality and bioprocess control, agriculture, biodefense and medical applications. Nanobiosensors application in the agricultural area has significantly improved productivity. The economic and cutting-edge nanobiosensors have been emphasised in this book chapter to address the difficulties in the main agricultural industry and emphasize the significance of nanobiosensor to detect insecticides, herbicides, fertilizers and diseases for increasing crop yield.

Keywords: Agriculture, Biosensors, Nanotechnology, Nanoparticles, Nanobiosensors.

INTRODUCTION

Sustainable agriculture is important to accomplish the goal of sustainable development of the United Nations. Nanotechnology has demonstrated remarkable possibilities for promoting sustainable agriculture. The Food and Agriculture Organization (FAO, 2017) estimates the need for food by 50%,

* **Corresponding author Tejveer Singh:** Translational Oncology Laboratory, Department of Zoology, Hansraj College, University of Delhi, New Delhi, Delhi-110067, India; E-mail: Tej6875@gmail.com

particularly in emerging nations, to demolish the hunger of the world's population, which will reach 10 billion by 2050 [1].

Additionally, there are 815 million people who are malnourished presently and by 2050, it is predicted that the count will increase to 2 billion. The global agricultural systems need to be drastically changing in light of this circumstance. Recent studies have demonstrated the great future of nanotechnology to enhance the agricultural section by boosting the efficacy of farm production and providing solutions to challenges in agriculture and the environment to increase food yield and sustainability. Therefore, global food security has become crucial, and the application of nanotechnology has achieved importance in recent times. However, the process of increasing food production has the hazardous impact on agricultural ecosystems, including the persistence of residual pesticide particles, the excess of heavy metals, and pollutants with harmful elemental particles. A wide range of health implications, including disorders of the bone marrow and nervous system, metabolic issues, infertility, disruption of cellular biological processes, as well as respiratory and immunological conditions, due to ingestion of such toxic elements into living beings through agricultural products.

Nano-biosensors are tiny element devices created to distinguish between certain molecules, biological elements or environmental factors. These sensors can detect at a level far lower than their macro scale analogs, are portable, cost-effective, and have a high degree of specificity. The industries of agriculture and food are a prime source of earnings and services for a sizeable percentage of the people. The agricultural sector contributes significantly to self-sustaining economic growth by giving humans the consumables they need and the raw materials for industrialization. The objective of a sensor is to identify changes in the environment, such as changes in temperature, humidity, water flow or light intensity, and to transmit the information to another electrical instrument that can analyze it. Nanobiosensors are characterized by their compactness, minimal rates, bio-compatibility, non-toxicity, portability, specificity, resilience, quick reaction times, absence of electrical noise, accuracy, precision, and reproducibility at optimal temperature and pH [2]. Nanobiosensors are used in agriculture to detect toxicants, pesticides, pollutants, veterinary medicines and heavy metals in food as well as direct and indirect food-borne pathogenic bacteria. Additionally, nanobiosensors can also identify antibiotic resistance, soil quality, crop stress and food quality [3]. In agriculture, nanobiosensors can be utilized to detect metal ions, a wide range of diseases, phytohormones, crop metabolites, insecticides, herbicides, pH and moisture of the soil. Nanobiosensors could enhance sustainable and healthy agriculture when it is used effectively with proper regulation and management. This chapter aims to provide a brief overview of the advancement and application of nanobiosensor in the field of agriculture.

BIOSENSOR

Biosensor is a tool that can identify various substances in biological or dietary samples, including environmental contaminants, vitamins, pesticide waste and biomolecules [4 - 9]. Biosensor is a tool of three components, including a bio-element, a transducer and a reading system (Fig. **1**). A bio-receptor is a biological living system like an organism, tissue and cell, or several biological components, including DNA/RNA, enzymes, receptors, peptide nucleic acids (PNA), antibodies and locked nucleic acids (LNA), organs, organelles, microbes, *etc.* make use of a biological mechanism for the recognition of analyte. The special interactions between the analyte and the receptor, such as enzymes-substrate (enzymatic interaction), microorganisms-proteins (cellular interaction), two complementary strands of nucleic acid (nucleic acid interactions) and antigen-antibody interaction [10 - 15], are converted into measurable effect by the transducers. Finally, such interactions are measured by a proper reading system (biosensor monitors) as shown in Fig. (**1**). These small devices are also classified according to how the signal transduction mechanism is used into many categories, including optical, electrochemical, piezoelectric, pyroelectric, electronic and gravimetric biosensors Many researchers are working to incorporate nanoparticles into biosensors construction to enhance the precision and functioning of a biosensor. Improvement in biosensor materials can play a significant role in the better application of sensing technology.

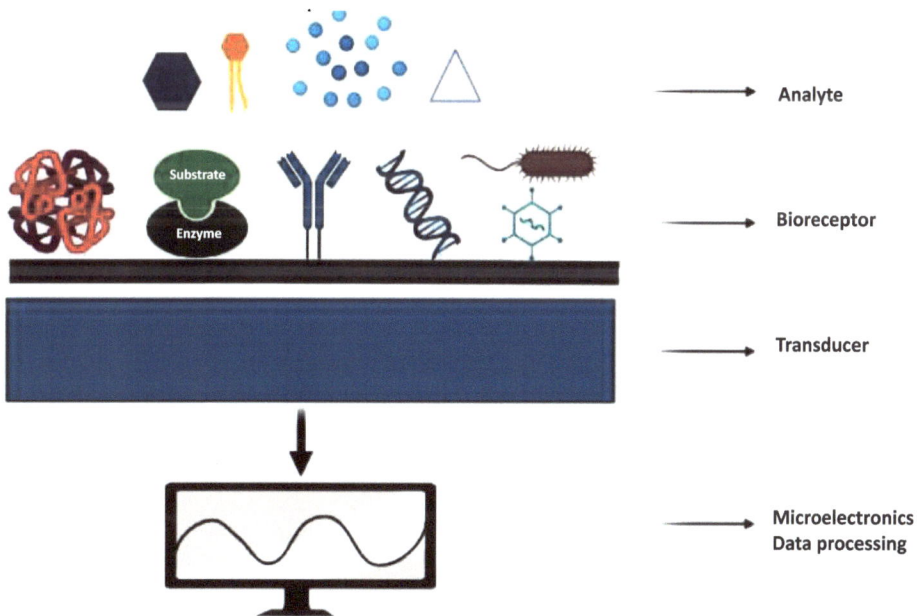

Fig. (1). Schematic diagram of the components of Biosensor.

NANOBIOSENSOR

The term "nanobiosensor" refers to a class of biosensors based on nanomaterials that have potential uses in medicine, the environment and agriculture. Nanobiosensors are modified biosensors with better detection specificity, selectivity and sensitivity (Fig. **2**). Nanobiosensors are also known as nanotechnology applied to biosensors which are connected to sensitized elements to identify specific analytes at ultra-low concentration *via* a physico-chemical transducer. An actual interface is a transducer to distinguish between signal occurrences in a digital signal. The detector finds signals from the microprocessor, receives the signal from the transducer, and sends it for amplification and analysis; the final output device display in terms of data. Some of the distinctive features of an ideal nanobiosensor are specificity, precision, affordable and quick approaches that can be automated and scaled (Fig. **1**), and they may even replace the traditional approaches [16]. In recent time, there is advancement in nanotechnology, with more reliable high-resolution sensors. Different branches of science like nanoscience, electronics, biology, computer and electronics are working together to generate biosensors with remarkable sensing potential. Nanobiosensor technology is useful in the speedy decision and early detection to increase the production of crops by appropriate execution of land, water, pesticides and fertilizer. This tool has advantageous characteristics over traditional sensors in features, such as a high ratio of surface-to-volume, conductivity, better performances, high precision and stability with prolonged life, rapid electron-transfer kinetics and expanded application in the detection of allergenic proteins [17]. The use of nanotechnology in interventions offers the stimulus to transform a variety of diagnostic fields, such as health, medicine, food, environment and the agricultural industry, transforming speculative properties into useful results [18 - 23]. Nanomaterials (NMs), like magnetic nanoparticles, gold nanoparticles, carbon nanotubes and quantum dots, have distinctive optical, chemical, physical, mechanical and magnetic features which are progressively used as biosensors and noticeably augment the recognition specificity and sensitivity. Nanobiosensor can identify a range of microorganisms, and several analytes, including urea, glucose, insecticides, *etc.* They can also be used for molecular analysis as well as to examine many metabolites.

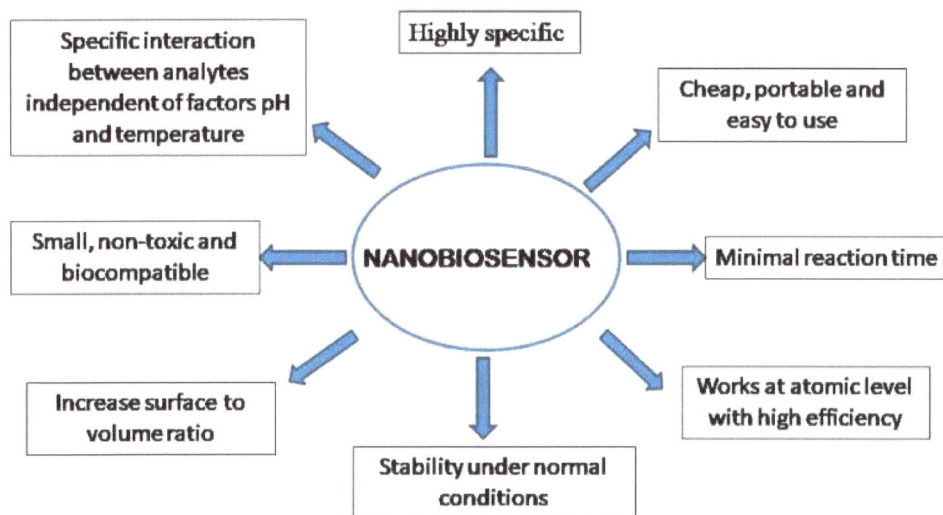

Fig. (2). Characteristic features of an ideal biosensor.

In the last decade, there are several promising NMs, such as graphene, nanocomposites, carbon nanotubes (CNTs), quantum dots (QDs), and nanoparticles (NPs), have been used for diagnostics and biosensors. The nanotechnology products can be classified into the following categories based on the number of dimensions "pushed" to the nanometer scale:

1. Thin films, such as coatings of implants for biocompatible purposes, coatings of pills, anticoagulant coatings of stents, and other therapeutic agents have only one dimension pushed to the scale of a few tens or hundreds of nanometres, while the other two dimensions can still extend up to millimetres;
2. Nanomaterials (NMs), such as silicon nanowires, nanorods, carbon nanotubes (CNTs) and fibres, have two dimensions pushed to the nanometer scale; and
3. Nanomaterials (NMs), such as gold, quantum dots, magnetic, liposomes and polymeric nanoparticles, have all three dimensions pushed to the nanometer scale.

Role of Nanobiosensor in Agriculture

Recent advancements in the creation of nanobiosensors have been made to enhance crop health through the detection of plant diseases, soil quality testing, pesticides and herbicides (Fig. **3**). The diagnostic technique for identifying plant pathogens uses nanoparticles [24]. There are enormous applications of nanomaterials in agriculture.

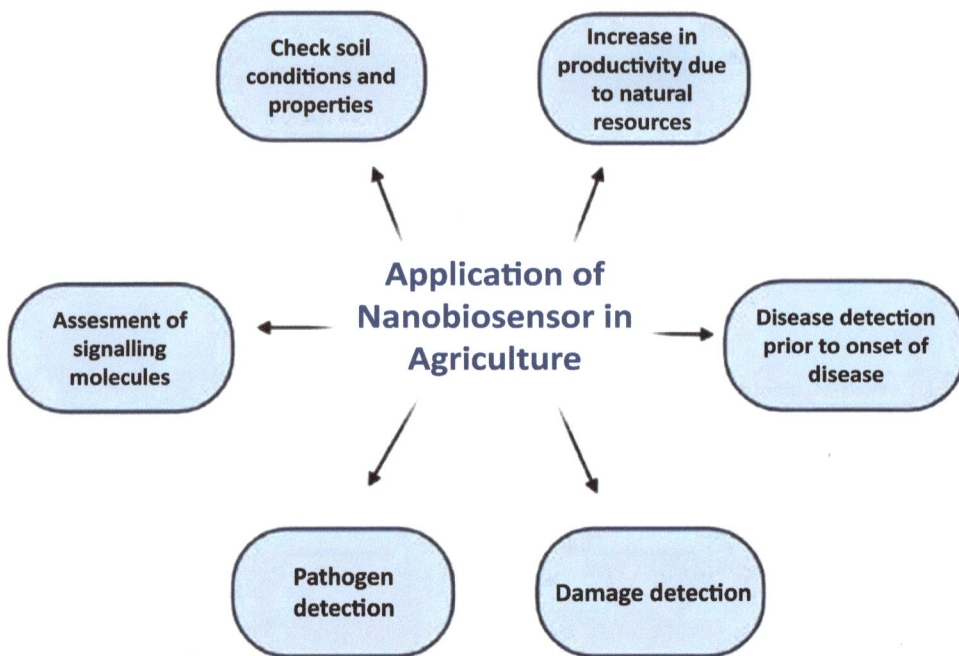

Fig. (3). Application of Nanobiosensor in Agriculture.

Detection of Pathogen with Nanobiosensors

In the food industry, laboratory-based techniques are fast and accurate to identify food pathogens-the first line of defence against outbreaks of foodborne illness. The development of nanobiosensors to identify pathogens, antibiotics, hormones, adulterants, and allergies in food may be made possible by the use of nanomaterials. *Vibrio cholera, Salmonella* spp., Campylobacter spp., *Escherichia coli* and *Clostridium spp.,* are the pathogens that pose a major risk to the animal's production and health, and they can result in gastroenteritis, food poisoning, *etc*. Due to contaminated meat and poultry and other sources of foodborne germs, salmonellosis in humans is a worldwide issue [25]. Food contaminated with *Salmonella* spp. could be detected using a nanobiosensor that had anti-*salmonella* polyclonal antibodies bound to streptavidin-biotin on the quantum dot surface [26]. Cholera toxin can be found using a liposomal poly(3,4-ethylenedioxythiophene)-coated carbon nanotube-based electrochemical immunosensor [27]. For fast monitoring of the food supply during storage, distribution and food processing, nanobiosensors are very promising. A nanobiosensor built on functionalized Fe_3O_4 nanoparticles and AuNPs associated monoclonal antibodies are used to detect *Escherichia coli* O157:H7 [28]. Detection of *Salmonella spp.* in skim milk, AuNPs-based nanobiosensors were also developed [29]. Mycotoxins are toxic substances made from fungi or moulds that are naturally occurring contaminants in food and animal feed products that are known to be nephrotoxic, hepatotoxic, mutagenic and carcinogenic, such as zearalenone, ochratoxin and aflatoxins. Due to these worries, it is urgently important to detect or keep an eye on foodborne pathogens as well as commercial feeds using a specialised tool called a nanobiosensor. Fluorescent oligo-capture probes detect the presence of a single nucleotide of plant-harmful bacteria and viruses on nanochips [30]. Investigation in the identification of the wheat kernel bunt disease pathogen using an immunosensor based on nanogold has been reported [31]. Numerous studies have revealed that it can identify plant pathogens, soil minerals, and viruses [32].

Detection of Drug and Residues

Nanobiosensors have the potential to recognize biological contaminants, such as veterinary drug residues in food, in addition to chemical and physical contaminants in food. They are used to assess and monitor the quality and safety of food. Antibiotics are commonly found in veterinary drugs, which are regularly administered to domesticated animals to fight diseases and encourage growth [33]. Antibiotic resistance to microbial diseases leads to epidemics in animals, which can pose serious risks to both people and animals [34, 35]. Milk can be detected with an aptamer biosensor [36]. Furthermore, aptamer-based array sensors and fluorescence biosensors can detect oxytetracycline and chloramphenicol [37, 38]. According to numerous studies, nanobiosensors can be used to find veterinary medicine residues in foods, including poultry products, milk and meat [38]. Nanobiosensors help to detect the presence of antibiotics effectively, reliably, and within a short period of time.

Detection of Pesticides

High dose of pesticide application in agriculture to protect plants increases toxicity to animals as well as to the environment. The traditional method of pesticide detection has many demerits results in the use of nanobiosensors which is more beneficial. Acetylcholinesterase nanobiosensor can be used for the detection of pesticides [39]. Detection of organophosphate and monocrotophos pesticides has been reported by an electrochemical biosensor [40]. Furthermore, using CdTe QDs and acetylcholinesterase, optical nanobiosensor can detect parathion and paraoxon pesticides [39]. Herbicide chlortoluron can be detected with the help of a nanobiosensor (inhibition of tyrosinase) [41]. Nanobiosensor was also used to detect the herbicides glufosinate, glyphosate, and their metabolites [42, 43]. Nanosmart dust can detect food contaminants and environmental pollutants [44]. Biosensor immobilized HRP on sulfonated polymer matrix was reported to detect the herbicide glyphosate [45]. In further work, AuNPs/(3-mercaptopropyl)-trimethoxysilane/Au electrode sensing surface was constructed for the electrochemical and biosensor methods of pesticide dichlorvos and carbamate detection. According to reports, an electrical circuit in the intelligent system examines and limits the usage of antibiotics and pesticides [46 - 49].

NANOBIOSENSOR AS A TOOL FOR PROMOTING SUSTAINABLE AGRICULTURE

The support of sustainable agriculture is greatly aided by nanofertilizers. Basically, a nano fertilizer is a substance that contains nanoparticles that improves crops and soil nutrients. Three methods can be used to create encapsulated nano fertilizer: isolating and dispersing nutrients as nanoscale emulsions or as individual particles; encapsulating nutrients in the form of nanotubes or nano-porous materials; and encapsulating nutrients by coating them with a thin protective polymer film [50 - 53].

Nanofertilizers are particularly effective at minimizing nitrogen loss [50, 54]. According to recent studies, ryegrass root tissues can be penetrated by zinc oxide nanoparticles, and carbon nanotubes can enter tomato seeds. Using the nanoscale porous domains on the surface of agricultural plants, it is suggested that a novel mechanism for the release and distribution of nutrients can be created. In this situation, nanofertilizers are only useful if they can regulate the release of nutrients at the appropriate moment and avoid the release of nutrients without any need that is not absorbed by the plant and transformed into gaseous form and contaminate the environment. In order to achieve this, biosensors enable the release of nutrients in response to demand and in accordance with soil nutrient conditions. Reduced negative effects of fertilizers are also made possible by controlled and delayed release.

Another illustration is zeolites, which are useful in improving crop output and soil quality. Zeolites, essentially aluminium silicate crystals, help enhance soil quality, plant growth and crop yield. They also increase water retention, infiltration and fertilizer effectiveness. They also help plants retain the soil's nutrients for a longer period of time and reduce the leaching of nutrients from the soil. Zeolite is thought to store nutrients in plant roots and deliver them to the plant when it needs them. It helps the plant utilize K and N fertilizers effectively, ensuring that no fertilizer is lost and that even a modest amount of fertilizer can result in a higher yield. Zeolite stays in the soil for a longer time and helps to improve the quality and duration of nutrient retention.

Zeolite and nanobiosensors together improved agriculture and significantly contributed to its development. Biosensors regulate the flow of nutrients or water that have been kept in zeolite when they detect a shortage within the soil or a plant. It is now possible to create insecticides that are enclosed in nanoparticles and emit pesticides over time and in response to environmental changes. In order to improve crop productivity, herbicides also have been paired with nanobiosensors that are given to vegetation only if necessary [55].

OTHER APPLICATIONS OF NANOBIOSENSORS

Nanobiosensors help in the measurement of the concentration of signaling molecules released by plants [56]. Phytoestrogens are synthesized by plants as defence mechanism against damage caused by fungi and are known as dietary estrogens. FRET probe using fluorescence signals and estrogen binding domain has been introduced to identify the phytoestrogens like resveratrol, genistein and daidzein. Nanobiosensors were also applied to detect dopamine level, which plays an important role in plant growth and development.

CONCLUDING REMARKS AND FUTURE PROSPECTS

Nanobiosensors can be employed starting from agri-based food, soil estimation, natural resources, estimation of pH, moisture of soil, disease regulation, recognition of pathogen, findings of chemicals and adulterants not suitable for humans to final commercialisation stage. Industries such as Nippon, Roche, IBM, *etc* deals with the manufacturing of nanobiosensors due to its wide application. Limited reports are accessible for commercialised nanobiosensors in agriculture, whereas few reports are available on commercial nanobiosensors in medical therapeutics, diagnosis and applications. The high cost involved in the manufacture, computerization assessment, results evaluation and justification of field trials that bring about the miniaturisation of prototypes into the industry for manufacture is still a major challenge. Agrifood nanotechnology aids our farmers by enhancing agricultural production and the agricultural industry through food manufacturing, storage and marketing. Nanomaterial-based biosensors and nanobiosensors are all being created to detect plant diseases and assess the soil's quality in order to improve the health of plants [57]. Nanotechnology, which primarily deals with the smallest particles is crucial in addressing agricultural issues that cannot be resolved using current methods. There are now more opportunities for usage across agriculture because of the creation of novel nanomaterials and nanodevices. One of the uses for nanotechnology is the creation of better biosensors, which leads to the creation of tiny structures called nanobiosensors that are more effective and well-organized than conventional biosensors [58]. By creating new nanomaterials, the use of biosensors and their efficacy can be further enhanced in the future, which will benefit the agriculture and food industries. As a result, nanotechnology's promise in the agri-food industry has not yet been fully uncovered. The combination of nanoscience, agro technology and biochemical engineering open up new possibilities in smart agriculture, sustainable development, mechanization, autonomous farming and cost-effective solutions. This technological advancement has significant effects on agriculture [58].

Nanobiosensor devices in the future have the potential to achieve smart agriculture with the GPS system. Such a strategy will help farmers to make decisions on farming, irrigation, pest control, fertilisation and harvesting with the usage of minimum natural resources. The customised nanobiosensors with high sensitivity and specificity will be realistic in the near future.

ACKNOWLEDGEMENT

Declared none.

REFERENCES

[1]   "The state of food and agriculture leveraging food systems for inclusive ruraltransformation", FAO: Rome, 2017.

[2]   R.P. Singh, J.W. Choi, A. Tiwari, and A.C. Pandey, "Functional nanomaterials for multifarious nanomedicine", In: *Biosensors Nanotechnology.,* A. Tiwari, A.P.F. Turner, Eds., Wiley, Inc.: Hoboken, NJ, USA, 2014.
        [http://dx.doi.org/10.1002/9781118773826.ch6]

[3]   R.P. Singh, "Nanobiosensors: Potentiality towards Bioanalysis", *J. Bioanal. Biomed.,* vol. 8, p. e143, 2016.
        [http://dx.doi.org/10.4172/1948-593X.1000e143]

[4]   S. Hassani, M. Rezaei Akmal, A. Salek Maghsoudi, S. Rahmani, F. Vakhshiteh, P. Norouzi, M.R. Ganjali, and M. Abdollahi, "High-performance voltammetricaptasensing platform for ultrasensitive detection of bisphenol A as an Environmental Pollutant", *Front. Bioeng. Biotechnol.,* vol. 8, p. 574846, 2020.
        [http://dx.doi.org/10.3389/fbioe.2020.574846] [PMID: 33015024]

[5]   N. Huang, M. Liu, H. Li, Y. Zhang, and S. Yao, "Synergetic signal amplification based on electrochemical reduced graphene oxide-ferrocene derivative hybrid and gold nanoparticles as an ultra-sensitive detection platform for bisphenol A", *Anal. Chim. Acta,* vol. 853, pp. 249-257, 2015.
        [http://dx.doi.org/10.1016/j.aca.2014.10.016] [PMID: 25467466]

[6]   D. Chauhan, R. Kumar, A.K. Panda, and P.R. Solanki, "An efficient electrochemical biosensor for Vitamin-D3 detection based on aspartic acid functionalized gadolinium oxide nanorods", *J. Mater. Res. Technol.,* vol. 8, no. 6, pp. 5490-5503, 2019.
        [http://dx.doi.org/10.1016/j.jmrt.2019.09.017]

[7]   G. Li, A. Wen, J. Liu, D. Wu, and Y. Wu, "Facile extraction and determination of organophosphorus pesticides in vegetables *via* magnetic functionalized covalent organic framework nanocomposites", *Food Chem.,* vol. 337, p. 127974, 2021.
        [http://dx.doi.org/10.1016/j.foodchem.2020.127974] [PMID: 32920274]

[8]   B. Bucur, F.D. Munteanu, J.L. Marty, and A. Vasilescu, "Advances in enzyme-based biosensors for pesticide detection", *Biosensors (Basel),* vol. 8, no. 2, p. 27, 2018.
        [http://dx.doi.org/10.3390/bios8020027] [PMID: 29565810]

[9]   S. Hassani, A. Salek Maghsoudi, M. Rezaei Akmal, S.R. Rahmani, P. Sarihi, M.R. Ganjali, P. Norouzi, and M. Abdollahi, "A sensitive aptamer-based biosensor for electrochemical quantification of PSA as a specific diagnostic marker of prostate cancer", *J. Pharm. Pharm. Sci.,* vol. 23, pp. 243-258, 2020.
[http://dx.doi.org/10.18433/jpps31171] [PMID: 32649855]

[10]   H Zhang, W Guan, L Zhang, X Guan, and S Wang, "Degradation of an organic dye by bisulfite catalytically activated with iron manganese oxides: The role of superoxide radicals", *ACS Omega.,* vol. 5, no. 29, pp. 18007-18012, 2020.
[http://dx.doi.org/10.1021/acsomega.0c01257] [PMID: 32743173]

[11]   X Chen, D Wang, T Wang, Z Yang, X Zou, and P Wang, "Enhanced photoresponsivity of a GaAs nanowire metalsemiconductor- metal photodetector by adjusting the Fermi level", *ACS Appl. Mater. Interfaces,* vol. 11, no. 36, pp. 33188-33193, 2019.
[http://dx.doi.org/10.1021/acsami.9b07891]

[12]   X Huang, Z Zeng, and H. Zhang, "Metal dichalcogenidenanosheets: preparation, properties and applications", *ChemSoc Rev,* vol. 42, p. 1934e46, 2013.

[13]   T Stephenson, Z Li, B Olsen, and D Mitlin, "Lithium ion battery applications of molybdenum disulfide (MoS2) nanocomposites", *Energy Environ Sci.,* vol. 7, p. 209e31, 2014.

[14]   T.N. Narayanan, C.S.R. Vusa, and S. Alwarappan, "Selective and efficient electrochemical biosensing of ultrathin molybdenum disulfide sheets", *Nanotechnology,* vol. 25, no. 33, p. 335702, 2014.
[http://dx.doi.org/10.1088/0957-4484/25/33/335702] [PMID: 25061018]

[15]   P. Damborský, J. Švitel, and J. Katrlík, "Optical biosensors", *Essays Biochem.,* vol. 60, no. 1, pp. 91-100, 2016.
[http://dx.doi.org/10.1042/EBC20150010] [PMID: 27365039]

[16]   R.C. Alves, M.F. Barroso, M.B. González-García, M.B.P.P. Oliveira, and C. Delerue-Matos, "New trends in food allergens detection: toward biosensing strategies", *Crit. Rev. Food Sci. Nutr.,* vol. 56, no. 14, pp. 2304-2319, 2016.
[http://dx.doi.org/10.1080/10408398.2013.831026] [PMID: 25779935]

[17]   V. Scognamiglio, "Nanotechnology in glucose monitoring: Advances and challenges in the last 10 years", *Biosens. Bioelectron.,* vol. 47, pp. 12-25, 2013.
[http://dx.doi.org/10.1016/j.bios.2013.02.043] [PMID: 23542065]

[18]   A.K. Srivastava, A. Dev, and S. Karmakar, "Nanosensors and nanobiosensors in food and agriculture", *Environ. Chem. Lett.,* vol. 16, no. 1, pp. 161-182, 2018.
[http://dx.doi.org/10.1007/s10311-017-0674-7]

[19]   M. Doroudian, A. O' Neill, R. Mac Loughlin, A. Prina-Mello, Y. Volkov, and S.C. Donnelly, "Nanotechnology in pulmonary medicine", *Curr. Opin. Pharmacol.,* vol. 56, pp. 85-92, 2021.
[http://dx.doi.org/10.1016/j.coph.2020.11.002] [PMID: 33341460]

[20]   S. Sahani, and Y.C. Sharma, "Advancements in applications of nanotechnology in global food industry", *Food Chem.,* vol. 342, p. 128318, 2021.
[http://dx.doi.org/10.1016/j.foodchem.2020.128318] [PMID: 33189478]

[21]   A. Acharya, and P.K. Pal, "Agriculture nanotechnology: Translating research outcome to field applications by influencing environmental sustainability", *NanoImpact,* vol. 19, p. 100232, 2020.
[http://dx.doi.org/10.1016/j.impact.2020.100232]

[22]   D. Dutta, and B.M. Das, "Scope of green nanotechnology towards amalgamation of green chemistry for cleaner environment: A review on synthesis and applications of green nanoparticles", *Environ. Nanotechnol. Monit. Manag.,* vol. 15, p. 100418, 2021.
[http://dx.doi.org/10.1016/j.enmm.2020.100418]

[23]   M. Usman, M. Farooq, A. Wakeel, A. Nawaz, S.A. Cheema, H. Rehman, I. Ashraf, and M. Sanaullah, "Nanotechnology in agriculture: Current status, challenges and future opportunities", *Sci. Total*

Environ., vol. 721, p. 137778, 2020.
[http://dx.doi.org/10.1016/j.scitotenv.2020.137778] [PMID: 32179352]

[24]   R.P. Singh, "Nanocomposites: recent trends, developments and applications", In: *Advances in Nanostructured Composites,* Nanotube Carbon, Composites Graphene, Eds., 1st Edition. CRC Press: Boca Raton, Florida, United States.. p. 552, 2018.

[25]   B. Nowak, T. von Müffling, S. Chaunchom, and J. Hartung, "*Salmonella* contamination in pigs at slaughter and on the farm: A field study using an antibody ELISA test and a PCR technique", *Int. J. Food Microbiol.,* vol. 115, no. 3, pp. 259-267, 2007.
[http://dx.doi.org/10.1016/j.ijfoodmicro.2006.10.045] [PMID: 17292500]

[26]   G. Kim, S.B. Park, J.H. Moon, and S. Lee, "Detection of pathogenic *Salmonella* with nanobiosensors", *Anal. Methods,* vol. 5, no. 20, pp. 5717-5723, 2013.
[http://dx.doi.org/10.1039/c3ay41351a]

[27]   S. Viswanathan, L. Wu, M.R. Huang, and J.A. Ho, "Electrochemical immunosensor for cholera toxin using liposomes and poly(3,4-ethylenedioxythiophene)-coated carbon nanotubes", *Anal. Chem.,* vol. 78, no. 4, pp. 1115-1121, 2006.
[http://dx.doi.org/10.1021/ac051435d] [PMID: 16478102]

[28]   Y. Wang, and E.C. Alocilja, "Gold nanoparticle-labeled biosensor for rapid and sensitive detection of bacterial pathogens", *J. Biol. Eng.,* vol. 9, no. 1, p. 16, 2015.
[http://dx.doi.org/10.1186/s13036-015-0014-z] [PMID: 26435738]

[29]   A.S. Afonso, B. Pérez-López, R.C. Faria, L.H.C. Mattoso, M. Hernández-Herrero, A.X. Roig-Sagués, M. Maltez-da Costa, and A. Merkoçi, "Electrochemical detection of *Salmonella* using gold nanoparticles", *Biosens. Bioelectron.,* vol. 40, no. 1, pp. 121-126, 2013.
[http://dx.doi.org/10.1016/j.bios.2012.06.054] [PMID: 22884647]

[30]   MM Lopez, P Llop, A Olmos, E Marco-Noales, M Cambra, and E Bertolini, "Are molecular tools solving the challenges posed by detection of plant pathogenic bacteria and viruses?", *Curr. Issues Mol. Biol.,* vol. 11, no. 1, pp. 13-46, 2009.
[PMID: 18577779]

[31]   KS Yao, SJ Li, KC Tzeng, TC Cheng, CY Chang, CY Chiu, CY Liao, J Hsu, and ZP Lin, "Fluorescence silica nanoprobe as a biomarker for rapid detection of plant pathogens", *Adv. Mater. Res.,* vol. 79, pp. 513-516, 2009.
[http://dx.doi.org/10.4028/www.scientific.net/AMR.79-82.513]

[32]   D.A. Brock, T.E. Douglas, D.C. Queller, and J.E. Strassmann, "Primitive agriculture in a social amoeba", *Nature,* vol. 469, no. 7330, pp. 393-396, 2011.
[http://dx.doi.org/10.1038/nature09668] [PMID: 21248849]

[33]   S.A. McEwen, and P.J. Fedorka-Cray, "Antimicrobial use and resistance in animals", *Clin. Infect. Dis.,* vol. 34, no. s3, suppl. Suppl. 3, pp. S93-S106, 2002.
[http://dx.doi.org/10.1086/340246] [PMID: 11988879]

[34]   S.B. Levy, and B. Marshall, "Antibacterial resistance worldwide: causes, challenges and responses", *Nat. Med.,* vol. 10, no. S12, suppl. Suppl., pp. S122-S129, 2004.
[http://dx.doi.org/10.1038/nm1145] [PMID: 15577930]

[35]   P. Courvalin, "Predictable and unpredictable evolution of antibiotic resistance", *J. Intern. Med.,* vol. 264, no. 1, pp. 4-16, 2008.
[http://dx.doi.org/10.1111/j.1365-2796.2008.01940.x] [PMID: 18397243]

[36]   K.M. Song, E. Jeong, W. Jeon, H. Jo, and C. Ban, "A coordination polymer nanobelt (CPNB)-based aptasensor for sulfadimethoxine", *Biosens. Bioelectron.,* vol. 33, no. 1, pp. 113-119, 2012.
[http://dx.doi.org/10.1016/j.bios.2011.12.034] [PMID: 22244734]

[37]   S. Wu, H. Zhang, Z. Shi, N. Duan, C. Fang, S. Dai, and Z. Wang, "Aptamer-based fluorescence biosensor for chloramphenicol determination using upconversion nanoparticles", *Food Control,* vol.

50, pp. 597-604, 2015.
[http://dx.doi.org/10.1016/j.foodcont.2014.10.003]

[38] H. Hou, X. Bai, C. Xing, N. Gu, B. Zhang, and J. Tang, "Aptamer-based cantilever array sensors for oxytetracycline detection", *Anal. Chem.,* vol. 85, no. 4, pp. 2010-2014, 2013.
[http://dx.doi.org/10.1021/ac3037574] [PMID: 23350586]

[39] S.P. Zhang, L.G. Shan, Z.R. Tian, Y. Zheng, L.Y. Shi, and D.S. Zhang, "Study of enzyme biosensor based on carbon nanotubes modified electrode for detection of pesticides residue", *Chin. Chem. Lett.,* vol. 19, no. 5, pp. 592-594, 2008.
[http://dx.doi.org/10.1016/j.cclet.2008.03.014]

[40] P. Norouzi, M. Pirali-Hamedani, M.R. Ganjal, and F. Faridbod, "A novel acetylcholinesterase biosensor for determination of monocrotophos using FFT continuous cyclic voltammetry", *Int. J. Electrochem. Sci.,* vol. 5, pp. 1434-1446, 2010.
[http://dx.doi.org/10.1016/S1452-3981(23)15370-3]

[41] M. Haddaoui, and N. Raouafi, "Chlortoluron-induced enzymatic activity inhibition in tyrosinase/ZnO NPs/SPCE biosensor for the detection of ppb levels of herbicide", *Sens. Actuators B Chem.,* vol. 219, pp. 171-178, 2015.
[http://dx.doi.org/10.1016/j.snb.2015.05.023]

[42] S.R. Mousavi, and M. Rezaei, "Nanotechnology in agriculture and food production", *J. Appl. Environ. Biol. Sci.,* vol. 1, no. 10, pp. 414-419, 2011.

[43] E.A. Songa, O.A. Arotiba, J.H.O. Owino, N. Jahed, P.G.L. Baker, and E.I. Iwuoha, "Electrochemical detection of glyphosate herbicide using horseradish peroxidase immobilized on sulfonated polymer matrix", *Bioelectrochemistry,* vol. 75, no. 2, pp. 117-123, 2009. a
[http://dx.doi.org/10.1016/j.bioelechem.2009.02.007] [PMID: 19336272]

[44] E.A. Songa, V.S. Somerset, T. Waryo, P.G.L. Baker, and E.I. Iwuoha, "Amperometric nanobiosensor for quantitative determination of glyphosate and glufosinate residues in corn samples", *Pure Appl. Chem.,* vol. 81, no. 1, pp. 123-139, 2009. b
[http://dx.doi.org/10.1351/PAC-CON-08-01-15]

[45] E.A. Songa, T. Waryo, N. Jahed, P.G.L. Baker, B.V. Kgarebe, and E.I. Iwuoha, "Electrochemical nanobiosensor for glyphosate herbicide and its metabolite", *Electroanalysis,* vol. 21, no. 3-5, pp. 671-674, 2009. c
[http://dx.doi.org/10.1002/elan.200804452]

[46] M. Sharon, A.K. Choudhary, and R. Kumar, "Nanotechnology in Agricultural Diseases", *J. Phytol.,* vol. 2, pp. 83-92, 2010.

[47] C.M. Monreal, M. De Rosa, S.C. Mallubhotla, P.S. Bindraban, and C. Dimkpa, "The application of nanotechnology for micronutrients in soil-plant systems. VFRC report", 2015.

[48] M. Kaushal, and S.P. Wani, "Nanobiosensors: frontiers in precision agriculture", In: *Nanotechnol.,* R. Prasad, Ed., Springer Nature Singapore Pte Ltd. 279-291, 2017.
[http://dx.doi.org/10.1007/978-981-10-4573-8_13]

[49] R.J. Chesterfield, J.H. Whitfield, B. Pouvreau, D. Cao, K. Alexandrov, C.A. Beveridge, and C.E. Vickers, "Rational design of novel fluorescent enzyme biosensors for direct detection of strigolactones", *ACS Synth. Biol.,* vol. 9, no. 8, pp. 2107-2118, 2020.
[http://dx.doi.org/10.1021/acssynbio.0c00192] [PMID: 32786922]

[50] P. N, S.S. M, B. A S, R.K. R, D. K P B, and C.S. B, "Bio-assisted synthesis of ferric sulphide nanoparticles for agricultural applications", *Kongunadu Research Journal,* vol. 7, no. 1, pp. 35-38, 2020.
[http://dx.doi.org/10.26524/krj.2020.6]

[51] R. Kalpana Sastry, S. Anshul, and N.H. Rao, "Nanotechnology in food processing sector-An assessment of emerging trends", *J. Food Sci. Technol.,* vol. 50, no. 5, pp. 831-841, 2013.

[http://dx.doi.org/10.1007/s13197-012-0873-y] [PMID: 24425990]

[52]   J. Chen, and B. Park, "Recent advancements in nanobioassays and nanobiosensors for foodborne pathogenic bacteria detection", *J. Food Prot.,* vol. 79, no. 6, pp. 1055-1069, 2016.
[http://dx.doi.org/10.4315/0362-028X.JFP-15-516] [PMID: 27296612]

[53]   M. Noruzi, "Electrospun nanofibres in agriculture and the food industry: a review", *J. Sci. Food Agric.,* vol. 96, no. 14, pp. 4663-4678, 2016.
[http://dx.doi.org/10.1002/jsfa.7737] [PMID: 27029997]

[54]   M.C. DeRosa, C. Monreal, M. Schnitzer, R. Walsh, and Y. Sultan, "Nanotechnology in fertilizers", *Nat. Nanotechnol.,* vol. 5, no. 2, p. 91, 2010.
[http://dx.doi.org/10.1038/nnano.2010.2] [PMID: 20130583]

[55]   D. Lin, and B. Xing, "Root uptake and phytotoxicity of ZnO nanoparticles", *Environ. Sci. Technol.,* vol. 42, no. 15, pp. 5580-5585, 2008.
[http://dx.doi.org/10.1021/es800422x] [PMID: 18754479]

[56]   S.Y. Kwak, M.H. Wong, T.T.S. Lew, G. Bisker, M.A. Lee, A. Kaplan, J. Dong, A.T. Liu, V.B. Koman, R. Sinclair, C. Hamann, and M.S. Strano, "Nanosensor technology applied to living plant systems", *Annu. Rev. Anal. Chem. (Palo Alto, Calif.),* vol. 10, no. 1, pp. 113-140, 2017.
[http://dx.doi.org/10.1146/annurev-anchem-061516-045310] [PMID: 28605605]

[57]   R.P. Singh, "Recent trends, prospects, and challenges of nanobiosensors in agriculture", In: *Biose. in agricul: recent trends and future perspect.* Springer: Cham. 3-13, 2021.

[58]   S. Mathivanan, "Perspectives of nano-materials and nanobiosensors in food dafety and agriculture", (ed. K. Krishnamoorthy), Novel Nanomaterials, Intech Open, London, 2021.
[http://dx.doi.org/10.5772/intechopen.95345]

SUBJECT INDEX

Y

Yellow fever 2

www.ingramcontent.com/pod-product-compliance
Lightning Source LLC
Chambersburg PA
CBHW041704210326
41598CB00007B/524